D0421284

FUNDAMENTAL ANALOG ELECTRONIC

FUNDAMENTAL ANALOG ELECTRONICS

Brian Lawless

Dublin City University

PRENTICE HALL
London New York Toronto Sydney Tokyo Singapore
Madrid Mexico City Munich

First published 1997 by
Prentice Hall Europe
Campus 400, Maylands Avenue
Hemel Hempstead
Hertfordshire, HP2 7EZ
A division of
Simon & Schuster International Group

© Prentice Hall Europe 1997

11604948

Learning Resources
Centre

All rights reserved. No part of this publication may be
reproduced, stored in a retrieval system, or transmitted,
in any form, or by any means, electronic, mechanical,
photocopying, recording or otherwise, without prior
permission, in writing, from the publisher.

Printed and bound in Great Britain by T.J.Press, (Padstow) Ltd.,
Padstow, Cornwall

Library of Congress Cataloging-in-Publication Data

Lawless, Brian
    Fundamental analog electronics: a one semester foundation
course / Brian Lawless
        p.   cm.
    Includes index.
    ISBN 0-13-534298-8
    1. Analog electronic systems. 2. Electronic circuits. 3. Title.
TK7867.L38   1996
621.381--dc20   96-6774
                                                     CIP

British Library Cataloguing-in-Publication Data

A catalogue record for this book is available
from the British Library

ISBN 0-13-534298-8

---

1 2 3 4 5   01 00 99 98 97

Contents

Lecturer's Preface

The widespread introduction of semesterized course structures has led to an increased combination of pressures on lecturers to cover material in a shorter teaching time and on students to gain an understanding of the material without any significant gap for study between the end of the course and the start of the examinations. The result is that material which does not conform to the strict objectives of the syllabus has to be somewhat curtailed and also that the texts used have to be more concise in their coverage of the core material. This text has been written in order to provide a foundation course in electronics which satisfies these constraints.

In structuring a course on a subject such as electronics, there are two possible approaches; the bottom-up, starting from components and working up through more complex combinations to the understanding of systems or the top-down, where the system is treated first and the details emerge at the later stages.

This writer's inclination is towards the first reductionist approach as experience has shown that progression through a sequence of fully understood elements gives the student a better grasp of the subject.

Within this bottom-up approach there is another division. The material can be presented so as to give an operational understanding in simple, but not simplistic, terms. Examples of this approach are the treatment of Fourier series in Unit 18 and the rules for the analysis of op-amps in Unit 39. Alternatively, the material can be presented with a full mathematical analysis in the first coverage. This second approach has been found to cause problems as the student is simultaneously trying to come to terms with unfamiliar mathematical methods and also trying to develop the intuitive understanding of circuits which is necessary for day-to-day work with electronics either as a user of existing circuits or as a developer of new circuits or applications.

Both the operational understanding and the mathematical understanding are necessary but they cannot be acquired simultaneously. The fully mathematical presentation is further complicated by the availability of powerful circuit simulation programs such as PSPICE in which the mathematical model is buried within the program and can only be extracted with difficulty. The danger of such programs is that they may come to replace the mathematical analysis of circuits.

This text is therefore a foundation course in analog electronics which is firmly rooted in a laboratory and practical environment and which treats material which can be covered in one semester.

Student's Preface

Have you ever watched a modern building being constructed? First the foundations are dug in a very muddy site, the steel frame is erected, the concrete floors are poured, the cladding walls, windows and interior walls are put in place. Finally the wiring, the gas and water pipes, the telephone cables are installed and after painting, the building is handed over by the builders. A long process, but it all started from a muddy hole in the ground.

Something very similar happens when a new subject is started, whether it is electronics, thermodynamics or French. There is an initial unorganized muddy field of unknowing and gradually order is imposed and a structure emerges. But the foundations have to be laid first and then a framework constructed. Once the framework is in position and understood, the finer points of the subject can be investigated in full and rigorous detail.

This text lays the foundations for an understanding of electronics. The coverage of the material is kept as straightforward as possible and excessive detail is avoided so as to get across the key points without becoming bogged down in material that can and must be added at a later time.

The more conventional and longer chapter structure, with each chapter covering a significant amount of material, has been avoided as chapters tend to contain more material than can be assimilated in one study session. Instead, the shorter unit structure contains an amount of material which can be covered in one session. Some of the units are very short but the time spent considering them should not be reduced as the simplest ideas are often the most important.

The structure of a succinct statement followed by worked examples and problems has been found to give the best results. Some of the problems have been developed to extend the material and also to show how electronics links into many other fields both in terms of concepts and also in terms of applications. These open ended problems may not be as readily solvable as the more numerical problems.

The summary section at the start of each unit should provide an aid in revision for students with an imminent examination. Asking the question 'Can I expand on the material of the summary?' is a valuable revision technique.

Finally, it is to be hoped that the reader will come to see and appreciate the intellectual creativity that has gone into the development of modern analog electronics, will learn to appreciate the rich literature in the field and possibly go on to make his or her own contribution to the subject.

Unit 1   Ohm's law

- The voltage (volts) measured across a resistor is equal to the current (amps) through the resistor multiplied by the resistance (ohms).

$$V = I \times R$$

Ohm's law was first stated by Georg Simon Ohm, in a pamphlet entitled *Die galvanische Kette, mathematisch bearbeitet* (The galvanic circuit investigated mathematically) which was published in 1827 in Cologne, Germany. As originally stated, the law related the voltage across a metallic conductor to the current through the conductor by a relationship, $V = I \times R$, where R is the resistance of the metallic conductor. Ohm's law, even as applied to metallic conductors, is an idealization, since the resistance of metals depends on the temperature. When Ohm's law is applied to nonmetallic conductors, the value of R is frequently not constant and as applied to structured devices such as diodes the deviations are such that the law can only be considered to apply over very small regions.

Electronic components which obey Ohm's law are so useful that significant effort has been expended in developing resistors which obey the law as closely as possible and which, when used in electronic circuits, will operate in a fully predictable manner. The result is that Ohm's law is probably the most important law in electronics and is certainly the most used.

Given any two of the three variables, the third value can be calculated immediately so, in trying to analyze a circuit, you should always concentrate on any resistor for which you can obtain or deduce two of the three terms used in Ohm's law.

Most modern discrete resistors are made using a metal film deposition process with the resistor encapsulated in a robust housing with connecting

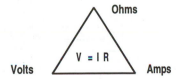

Figure 1.1: Given any two quantities, the third is determined.

wires; hence the circuit symbol of a rectangle shown in Figure 1.2. However, in the early days of electronics, the resistors were made from thin wires wound in a zig-zag pattern on a former and this led to the representation of resistors by the second symbol shown in the diagram. We use this symbol in the text since it is preferred by scientists and engineers because it is quicker to sketch. Being able to draw circuits quickly is an important aspect of learning electronics as it helps you to recognize and understand the shapes of circuits. When the circuit finally works, then you can use the formal drawing office representation of resistors for the archival drawing.

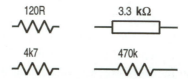

Figure 1.2: Circuit symbols for resistors.

On the circuit diagram, the value of the resistor is printed beside the symbol as 120 R, 3.3 kΩ, 4k7, 470 k. One convention is to put the multiplier, R, k or M, in the position of the decimal point which helps to avoid any ambiguity between $4.7\,k\Omega$ and $47\,k\Omega$ due to the decimal point becoming indistinct on a worn diagram. In many circuit diagrams you will find that the Ω symbol identifying the units as ohms is left out as it is considered obvious that the number beside a resistance represents units of ohms.

Resistors are usually too small for the value of the resistance to be printed on the component so the values are denoted by coloured bands on the component with the colours representing the numerical digits. This is shown in Figure 1.3. Beware of one problem: about 5% of men are colour blind and they should take care to use new resistors out of marked packages or else to check the value of the resistors with a multimeter before using them.

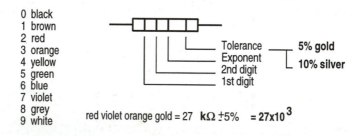

Figure 1.3: Resistor colour code marking method.

When resistors are manufactured, there is a tolerance on the actual values of the resistors. The typical resistor used in electronics has a tolerance of ±5%. This has one important implication for anyone learning electronics. There is no point trying to calculate resistor values to a percentage accuracy greater than 5% since the components are only rated at this accuracy. If you need resistors having an accuracy better than 5% in a circuit then you should usually build in trimmer resistors which can be adjusted to tune the circuit after it is assembled.

If you look at the rack of drawers containing resistors in the lab or the range of resistors which are available from the electronic component suppliers, then you will find that resistors come in strange numerical values which are multiples of 10 of the values 10, 12, 15, 18, 22, 27, 33 etc. This gives the set of preferred values for resistors. In the early days the manufacturers made the resistors and then classified them. By using successive nominal values which were 120% of the lower value it was possible to classify any resistor into a nominal value ±10% and thus sell any resistor made! Even though resistor manufacturing technology has improved, the same set of preferred values still remain in use. The other more relevant advantage of these values is that when you design a circuit and calculate the resistor values, you will always be able to find a preferred value within 5% or 10% of whatever value you calculate.

To return to Ohm's law, $V = I \times R$, the crucial thing to remember is that we use the voltage difference between one end of the resistor and the other—visualize a voltmeter connected across the resistor. We also use the current flowing through the resistor. These points will be illustrated in some of the examples.

In specifying a voltage the voltage measurement is usually quoted as either a voltage difference across a component such as a resistor or a voltage difference from a point to a reference common or ground point. In most cases we will assume that the reference common or ground is the reference line at the bottom of the circuit diagram as shown in Figure 1.4 (a). In a more

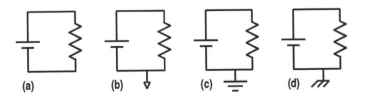

Figure 1.4: Common, ground and earth symbols.

complex circuit, where a single ground line in the diagram would unduly complicate the diagram, the reference ground may be indicated by an arrow as shown in Figure 1.4 (b) and connections may be made to ground from any point in the circuit diagram by use of this symbol. When it is necessary to make a connection to ground for safety reasons such as the ground connection in a three pin mains plug, then this is indicated by the symbol in Figure 1.4 (c). In situations where a very substantial ground may be needed, such as the ground in a radio transmitter antenna, then the symbol in Figure 1.4 (d) may be used. This graduation of grounding efficiency from local reference to substantial amounts of buried metallic conductors may not always be of significance in all circuits encountered but the possibility should be borne in mind.

1.1 Examples

1.1 If a current of 1.2 A flows in the resistor in the circuit in Figure 1.5, calculate the voltage drop across the resistor.

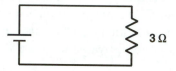

Figure 1.5: Example 1.1.

Voltage across resistor is $V = I \times R = 1.2\,\text{A} \times 3\,\Omega = 3.6\,\text{V}$

1.2 In the circuit in Figure 1.6, if the voltage across the $17\,\Omega$ resistor is measured to be 34.0 V, calculate the current in the resistor.

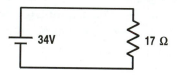

Figure 1.6: Example 1.2.

The current, $I = \frac{V}{R} = \frac{34.0\,\text{V}}{17\,\Omega} = 2.0\,\text{A}$.

1.2 Problems

1.1 If a current of 3.2×10^{-6} A flows through a $4.7 \times 10^{3}\,\Omega$ or $4.7\,\text{k}\Omega$ resistor, calculate the voltage across the resistor.

1.2 If the voltage across a resistor is 23 V and the current flowing is 1.9 mA or 1.9×10^{-3} A, calculate the value of the resistance.

1.3 What are the maximum and minimum values for a 5% tolerance 68 kΩ resistor? If the voltage across this 68 kΩ resistor is measured to be 21.3 V, calculate the nominal, the maximum and the minimum values of the current flowing through the resistor.

1.4 If the voltage at one end of a 4.7 kΩ resistor, measured with respect to (wrt) ground, is +49 V and the voltage at the other end, measured wrt ground, is +58 V, calculate the voltage difference or potential difference across the resistor.
Calculate the current flowing through the resistor.
A possible circuit is shown in Figure 1.7.
Where would you connect a voltmeter in order to measure the voltage across the resistor?

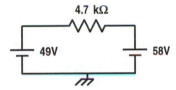

Figure 1.7: Problem 1.4.

1.5 Redraw the circuit shown in Figure 1.8 and show where you would connect a current meter so as to measure the current in the 4.7 kΩ resistor. Use a circle with an A in it to represent the current meter. What value of current would you expect to obtain?

Figure 1.8: Problem 1.5.

1.6 What is the resistance of a digital multimeter set to measure volts?

1.7 What is the resistance of a digital multimeter set to measure amps?

1.8 What are the colour code markings on resistors having the following values: $22\,k\Omega\pm5\%$, $1.5\,M\Omega\pm10\%$, $47\,k\Omega\pm5\%$, $560\,k\Omega\pm10\%$, $68\,\Omega\pm5\%$?

1.9 Redraw the circuit shown in Figure 1.9 and show where you would connect a voltmeter to measure the voltage across the $1.8\,k\Omega$ resistor. Use a circle with a V in it to represent the voltmeter.

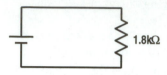

Figure 1.9: Problem 1.9.

1.10 What are the values of resistors having these markings?

Band 1	Band 2	Band 3	Band 4	Value
Orange	White	Red	Gold	......
Orange	Orange	Orange	Gold	......
Violet	Green	Brown	Gold	......
Brown	Black	Brown	Silver	......
Grey	Red	Yellow	Gold	......

1.11 The resistivity of nichrome alloy is $1.0\times10^{-6}\,\Omega m$. A $4.7\,k\Omega$ resistor is to be fabricated as a 12 mm long, $10\,\mu m$ wide track of nichrome film on an insulating substrate. Calculate the required thickness of the nichrome film.

Unit 2   Resistors in series

- When resistors are connected in series, the same current flows in all of the resistors.

- The voltage drop across the equivalent resistor R_{series} is:

$$V_{series} = IR_{series} = IR_1 + IR_2 + IR_3 + \cdots$$

Therefore $R_{series} = R_1 + R_2 + R_3 + \cdots$

Figure 2.1: Resistors in series.

An electric current is due essentially to the movement of electrons through the conducting wires and resistors. If the wires and resistors are connected in series, there is no alternative path for the electrons. There can be no accumulation of electrons in a resistor just as there can be no accumulation of water in a pipe carrying a flow of water. Any electrons entering one end of a set of resistors, wired in series, are matched by an equal number of electrons emerging from the other end. There is a slight delay, however, before the electrons emerge from the other end which is given approximately by:

$$\text{Delay} \approx \frac{\text{Distance between ends}}{\text{Speed of light}}$$

where speed of light $= c = 3 \times 10^8 \, \text{m s}^{-1}$.

For a circuit extending over $0.3 \, \text{m}$ or one foot (the typical size of a box of electronics) the delay will be approximately $1 \, \text{ns} = 10^{-9} \, \text{s}$ which is negligible unless you are dealing with very fast digital electronics or radio frequency circuits. Microcomputers are beginning to approach these switching times and the physical size of the circuit boards can be a problem— hence the drive towards smaller and more densely packed integrated circuits and surface mounted devices on printed circuit boards. There are other good reasons for having physically small electronic circuits but large and fast electronic circuits are not possible without using travelling wave circuit technology.

2.1   Examples

2.1 Calculate the current in the circuit of Figure 2.2, given that the voltage across the two resistors in series is 3.8 V. Calculate the current in the 100 R resistor.

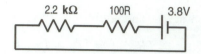

Figure 2.2: Example 2.1.

$$R_s = R_1 + R_2 = 2.2\,\text{k} + 100 = 2.2 \times 10^3 + 100 = 2300\,\Omega$$

$$I = \frac{3.8\,\text{V}}{2300\,\Omega} = 1.65 \times 10^{-3} = 1.65\,\text{mA}$$

2.2 Calculate the current in the circuit of Figure 2.3, given that the voltage across the 1.8 kΩ resistor is 6.4 V. Calculate the voltage across the 2.7 kΩ resistor. Calculate the battery voltage.

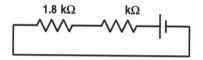

Figure 2.3: Example 2.2.

$$\text{Current} = I = \frac{6.4\,\text{V}}{1.8\,\text{k}\Omega} = \frac{6.4}{1800} = 3.55 \times 10^{-3} = 3.55\,\text{mA}$$

$$V_{2.7\,\text{k}\Omega} = 3.55\,\text{mA} \times 2.7\,\text{k}\Omega = 3.55 \times 10^{-3} \times 2.7 \times 10^3 = 9.6\,\text{V}$$

$$V_{Battery} = 6.4\,\text{V} + 9.6\,\text{V} = 16\,\text{V}$$

2.2   Problems

2.1 If the current is measured to be 2.3 mA, calculate the voltage across each of the resistors in the circuit in Figure 2.4. What is the total voltage across the resistors in series? Where would you insert the current meter into the circuit in order to measure the current? What current range setting would you use on the digital multimeter?

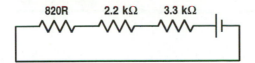

Figure 2.4: Problem 2.1.

2.2 In the circuit of Figure 2.5, a voltage of 0.36 V is measured at point B relative to ground. What is the current in the resistors? What is the voltage drop across the 4.7 kΩ resistor? What is the voltage at point A relative to ground?

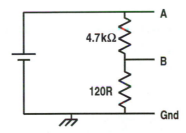

Figure 2.5: Problem 2.2.

2.3 Calculate the voltages at points A and B in the circuit of Figure 2.6.

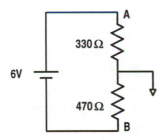

Figure 2.6: Problem 2.3.

2.4 If a digital multimeter, set to measure current, is inserted into a circuit such as that in Figure 2.4 it will be found that the current reading is not the same on all range settings. Explain why this is so.

Unit 3   Resistors in parallel

- When resistors are connected in parallel, the same voltage difference or potential difference is present across all of the resistors.

$$\text{That is} \quad V_p = I_{Total}R_p = I_1 R_1 = I_2 R_2 = I_3 R_3 = \cdots$$

$$\text{but} \quad I_{Total} = I_1 + I_2 + I_3 + \cdots$$

$$\text{Therefore} \quad \frac{1}{R_p} = \frac{1}{R_1} + \frac{1}{R_2} + \frac{1}{R_3} + \cdots$$

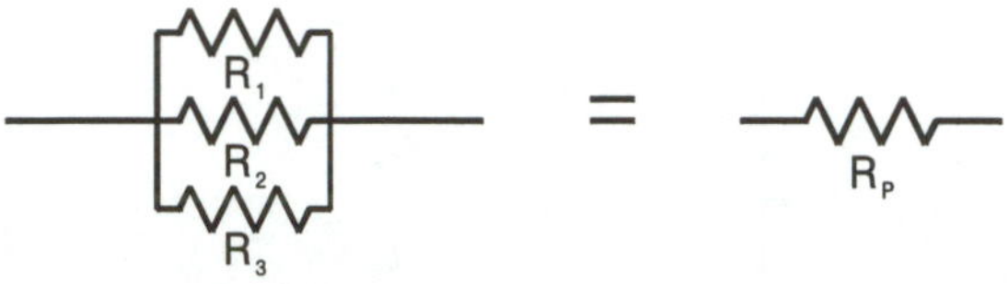

Figure 3.1: Resistors in parallel.

When current, flowing through a circuit, meets a number of resistors connected in parallel, the current divides between the resistors with part of the current flowing in each resistor. The sum of the currents in each of the resistors equals the total current flowing around the circuit. The voltage difference between the point where the current divides and the point where the current recombines is the same for all of the resistors and therefore the voltage difference across all of the resistors is the same. This gives the two equations in the summary: the IR value for each of the resistors is equal to the voltage across the resistors and the total current is the sum of the currents in the individual resistors. We can therefore consider a single resistor which we call R_p and is the equivalent parallel resistor which will allow the same total current to flow with the same voltage across the resistor as the n resistors connected in parallel.

A good analogy is to consider a river flowing past a number of islands. The total flow divides up into a number of smaller streams in the channels between the islands and recombines downstream. The change in level or head loss between a point upstream of the islands and a point downstream of the islands is the same for all possible routes taken by a cork floating in the stream.

3.1 Examples

3.1 Calculate the equivalent resistance for the parallel resistor circuit shown in Figure 3.2.

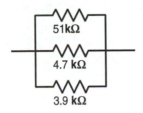

Figure 3.2: Example 3.1.

$$\frac{1}{R_p} = \frac{1}{51\,\text{k}\Omega} + \frac{1}{4.7\,\text{k}\Omega} + \frac{1}{3.9\,\text{k}\Omega}$$
$$= 4.89 \times 10^{-4}\,\Omega^{-1}$$
$$\text{Therefore} \quad R_p = 2046\,\Omega$$
$$= 2.046\,\text{k}\Omega$$

3.2 In the circuit shown in Figure 3.3, the current in the $820\,\Omega$ resistor is measured to be $2.5\,\text{mA}$. Calculate the battery voltage, V_B, and also calculate the total current flowing through the battery. Calculate the current in the $3.9\,\text{k}\Omega$ resistor.

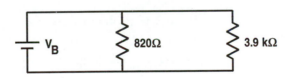

Figure 3.3: Example 3.2.

$$V_B = 2.5\,\text{mA} \times 820\,\Omega$$
$$= 2.5 \times 10^{-3} \times 820\,\text{V}$$
$$= 2.05\,\text{V}$$

$$\frac{1}{R_p} = \frac{1}{820} + \frac{1}{3900}$$
$$= 1.48 \times 10^{-3}\,\Omega^{-1}$$

$$\text{Therefore} \quad R_p = 678\,\Omega$$
$$\text{Hence} \quad I_{Total} = \frac{2.05\,\text{V}}{678\,\Omega}$$
$$= 3.03 \times 10^{-3}\,\text{A}$$
$$= 3.03\,\text{mA}$$
$$\text{and} \quad I_{3.9\,k\Omega} = \frac{2.05\,\text{V}}{3.9 \times 10^{3}\,\Omega}$$
$$= 5.26 \times 10^{-4}\,\text{A}$$
$$= 0.526\,\text{mA} \quad \text{or} \quad 526\,\mu\text{A}$$

3.2 Problems

3.1 Use the equation for the potential difference and the equation for the total current stated in the summary to derive the equation for the equivalent parallel resistor stated in the summary.

3.2 Calculate the equivalent resistance for the parallel resistor circuit shown in Figure 3.4.

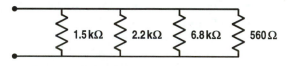

Figure 3.4: Problem 3.2.

3.3 If the current in the 470 Ω resistor in Figure 3.5 is measured to be 1.9 mA, calculate the battery voltage, V_B, and also calculate the total current flowing through the battery. Calculate the currents in the 5.6 kΩ and 6.8 kΩ resistors.

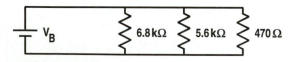

Figure 3.5: Problem 3.3.

3.4 If the total current flowing in the circuit in Figure 3.6 is measured to be 45 mA, calculate the voltage setting of the variable voltage power supply.

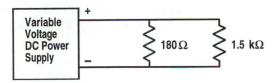

Figure 3.6: Problem 3.4.

3.5 In the circuit in Figure 3.7, the battery voltage is 10 V. Calculate the total current drawn from the battery. Calculate the current in each of the resistors.

Figure 3.7: Problem 3.5.

3.6 A value of $3.6\,\Omega$ was measured for the resistance of 90 m of 16/0.2 mm insulated equipment wire (16 strands each of 0.2 mm diameter). Calculate the cross sectional area of the wire. Calculate the resistance of 1 m of a single strand of the wire. (Insulated equipment wire of this cross section of copper conductor would typically be rated to carry a maximum current of 3 A without overheating.)

Unit 4   Potential divider

- The current in the resistors in series in a potential divider chain is given by the input voltage divided by the sum of the resistors in the chain.

$$\text{Current} \quad I = \frac{V_{in}}{R_1 + R_2}$$

- The output voltage is given by this current times the resistor across the output.

$$\text{Output voltage} \quad V_{out} = I \times R_2 = \frac{R_2}{R_1 + R_2} \times V_{in}$$

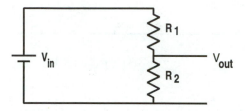

Figure 4.1: Potential divider.

When calculating the output from a potential divider circuit, it is better to obtain the current in the resistor chain first, by getting the equivalent series resistance. The output voltage is then obtained by using Ohm's law to calculate the voltage drop for the resistor across the output. This approach avoids the need to remember which is R_1 and which is R_2 with the consequent possibility of confusion.

Think of the output voltage as a fraction of the input voltage. The output voltage will therefore always be smaller than the input voltage but be careful when using bipolar supplies such as are used in Problem 4.4.

Frequently it is necessary to be able to switch between a set of fixed fractions of the input voltage such as when changing ranges in voltmeters or oscilloscopes. In this case, rotary switches or slide switches are used.

14

If a continuously variable fraction of the input voltage is required then a rotary or slider potentiometer is used where a wiper varies the contact point along a resistive track of carbon, wire or conductive plastic. Two types of track are used: a linear track where the resistance is proportional to angle of rotation or linear movement and a logarithmic track where the logarithm of the resistance is proportional to angle of rotation or linear movement. The response of the ear to increases in audio power is nonlinear and the perceived loudness of sound is proportional to the logarithm of the sound intensity or watts per square metre, Wm^{-2}. Therefore log pots are often used as volume control potentiometers to give an apparent linear increase in loudness with rotation of the volume control knob.

4.1   Examples

4.1 Calculate the output voltage from the potential divider in Figure 4.2.

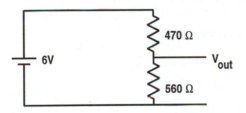

Figure 4.2: Example 4.1.

$$\text{Current} \quad I \;=\; \frac{6\,\text{V}}{470\,\Omega + 560\,\Omega}$$
$$=\; \frac{6}{1030}$$
$$=\; 5.82\,\text{mA}$$
$$\text{Output voltage} \quad V_{out} \;=\; 5.82\,\text{mA} \times 560\,\Omega$$
$$=\; 3.26\,\text{V}$$

4.2 Calculate the output voltage when the potentiometer shown in Figure 4.3 is set at 0%, 12%, 50%, 75%, 90%, 100% of its range.

$$\text{The percentage of range} \;=\; \frac{R_2}{R_1 + R_2} \times 100$$
$$\text{Therefore} \quad V_{out} \;=\; 0.00 \times 15\,\text{V} = 0.0\,\text{V} \quad \text{for 0\% setting}$$
$$V_{out} \;=\; 0.12 \times 15\,\text{V} = 1.8\,\text{V} \quad \text{for 12\% setting}$$

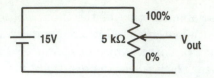

Figure 4.3: Example 4.2.

$$V_{out} = 0.50 \times 15\,\text{V} = 7.50\,\text{V} \quad \text{for 50\% setting}$$
$$V_{out} = 0.75 \times 15\,\text{V} = 11.3\,\text{V} \quad \text{for 75\% setting}$$
$$V_{out} = 0.90 \times 15\,\text{V} = 13.5\,\text{V} \quad \text{for 90\% setting}$$
$$V_{out} = 1.00 \times 15\,\text{V} = 15.0\,\text{V} \quad \text{for 100\% setting}$$

4.2 Problems

4.1 The rotary switch connects the output to one of the points 1 to 5 in the circuit of Figure 4.4. The total resistance of the series of resistors is to be 1 MΩ. Calculate the values of R_1, R_2, R_3 and R_4 such that the output voltages are given by:

$$V_1 = V_{in}, \ V_2 = \frac{V_{in}}{2}, \ V_3 = \frac{V_{in}}{10}, \ V_4 = \frac{V_{in}}{20}, \ V_5 = 0$$

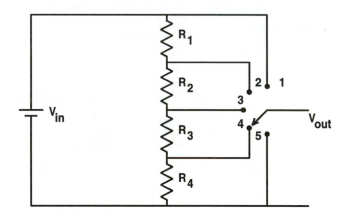

Figure 4.4: Problem 4.1.

4.2 Calculate the output voltage for each setting of the rotary switch shown in Figure 4.5. Why should a make-before-break switch be used, if the output voltage is never to exceed 2 V? Why does this difficulty not arise with the circuit configuration used in Problem 4.1?

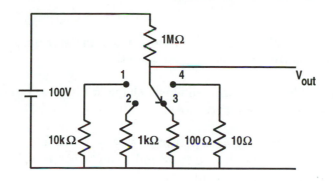

Figure 4.5: Problem 4.2.

4.3 Plot a graph of the output voltage as a function of the percentage range setting of the 1 kΩ potentiometer in Figure 4.6.

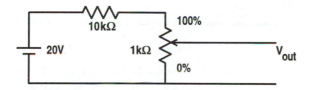

Figure 4.6: Problem 4.3.

4.4 Calculate the voltages, measured wrt ground, which you would expect to observe at points A, B, C, D, E and F of Figure 4.7. Follow the convention that ground is the line at the bottom of the circuit unless otherwise indicated. (Take care to note the polarity of the battery supplies.)

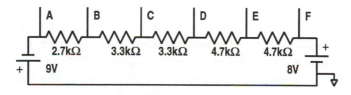

Figure 4.7: Problem 4.4.

Unit 5 Resistor networks

Very great simplifications can often be made in determining resistor networks if the following four procedures are utilized:

- Redraw the circuit in a different shape while maintaining the same connections or circuit topology.

- If, by symmetry arguments or otherwise, you can find nodes which are at the same potential, then these nodes can be connected and significant circuit simplifications made.

- For components connected in series, use the fact that the same current flows in all the components.

- For components connected in parallel, use the fact that the voltage across all the components is the same.

The currents and voltages in any network of resistors can be determined by the application of the principles and methods collectively known as Kirchhoff's laws. The difficulty is that significant computation, either hand calculation or computer programming, is often required which takes time and does not always give insight into the characteristics of the circuit. If the insights which come from understanding a circuit, as distinct from simply calculating the answer, are missing, then it is unlikely that new or original circuits will be developed.

Very often a circuit is encountered in texts or manuals which is unfamilar and whose performance is unknown. In such cases, you should try redrawing the circuit in a different shape so that the component connection or topology is maintained. It will often be found that the circuit is one with which you are already familiar. Another variant of this approach is to split a large, complex circuit into blocks, each having a known characteristic.

Many resistor networks, particularly those constructed from identically valued resistors, have axes of symmetry and therefore have nodes which must be at the same potential. If two nodes which are at the same potential are connected together, no current flows from node to node so there will be no change caused to the voltages in the circuit by making the connection. The change made to the circuit may, however, make significant simplifications to the analysis of the circuit.

In particular, Example 5.3 shows how a circuit can be analyzed by re-drawing the circuit rather than by formal numerical analysis.

5.1 Examples

5.1 Calculate the resistance between points A and B in Figure 5.1.

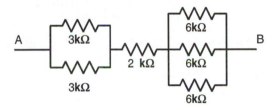

Figure 5.1: Example 5.1.

$$
\begin{aligned}
R_{AB} &= \frac{3\,\text{k}\Omega}{2} + 2\,\text{k}\Omega + \frac{6\,\text{k}\Omega}{3} \\
&= 1.5\,\text{k}\Omega + 2\,\text{k}\Omega + 2\,\text{k}\Omega \\
&= 5.5\,\text{k}\Omega
\end{aligned}
$$

5.2 Calculate the voltages at points A, B, C, D, E in the resistive ladder network shown in Figure 5.2.

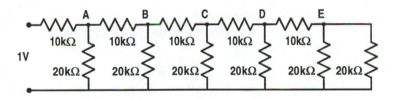

Figure 5.2: R–$2R$ resistive ladder.

This problem is derived from circuits used in the R–$2R$ type of digital to analog converters which we will meet later in Unit 54. The analysis is best carried out using a recursive technique as follows:

Consider two $10\,\text{k}\Omega$ resistors in series as shown in Figure 5.3 (a). Looking from the left they appear as a $20\,\text{k}\Omega$ resistor.

If one of the $10\,\text{k}\Omega$ resistors is replaced by two $20\,\text{k}\Omega$ resistors in parallel, we get the circuit in Figure 5.3 (b). This circuit still has a $20\,\text{k}\Omega$ input

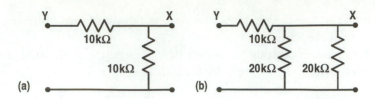

Figure 5.3: Equivalent circuits.

resistance. Also the voltage at point X is half of the voltage at point Y since we have effectively two equal resistors in a potential divider.

Now replace the right hand $20\,\text{k}\Omega$ resistor in Figure 5.3 (b) by a copy of the original circuit which has an input resistance of $20\,\text{k}\Omega$ as shown in Figure 5.4.

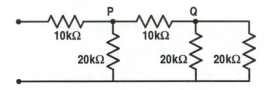

Figure 5.4: Make a substitution.

At point Q, the voltage is half of the voltage at point P because of the two $20\,\text{k}\Omega$ resistors in parallel acting as a $10\,\text{k}\Omega$ and forming a potential divider with the $10\,\text{k}\Omega$ between P and Q. At point P, the $20\,\text{k}\Omega$ and the $20\,\text{k}\Omega$ input resistance of the circuit to the right act as a $10\,\text{k}\Omega$ to form a potential divider so that the voltage at P is half of the voltage at the input to the circuit.

We can then see that the voltage at each node, A, B, C, D and E, in Figure 5.2 is half of the value at the earlier node. The voltages are therefore:

$$V_A = \frac{1}{2}\,\text{V}, \quad V_B = \frac{1}{4}\,\text{V}, \quad V_C = \frac{1}{8}\,\text{V}, \quad V_D = \frac{1}{16}\,\text{V}, \quad V_E = \frac{1}{32}\,\text{V}$$

5.3 Determine the resistance between nodes A and B in the resistor network in Figure 5.5, given that all of the resistors are of value $1\,\text{k}\Omega$.

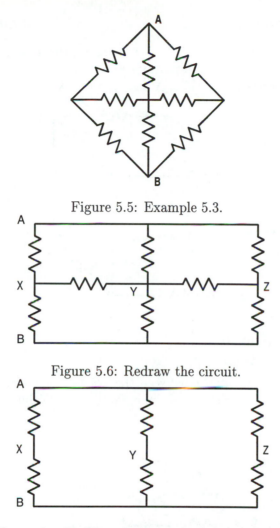

Figure 5.5: Example 5.3.

Figure 5.6: Redraw the circuit.

Figure 5.7: The nodes X, Y and Z are at the same potential, therefore replacing the cross resistors will not affect the circuit because no current will flow in the cross resistors. Therefore the cross resistors can be omitted.

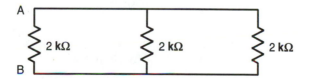

Figure 5.8: Combine series $1\,\mathrm{k}\Omega$ resistors and then combine the parallel $2\,\mathrm{k}\Omega$ resistors to obtain a single $666\,\Omega$ resistor.

5.2   Problems

5.1 Consider a cube having $1\,\mathrm{k\Omega}$ resistors along the 12 edges of the cube as shown in Figure 5.9. Determine the resistance between diagonally opposite corners of the cube, A and B, as shown in the diagram.

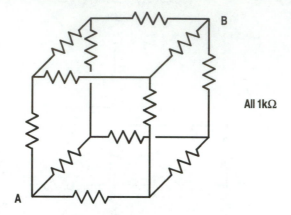

Figure 5.9: Problem 5.1.

5.2 Determine the resistance between the points A and B in the circuit in Figure 5.10, given that all of the resistors are of value $1\,\mathrm{k\Omega}$.

Figure 5.10: Problem 5.2.

5.3 Calculate the voltages at nodes A, B, C, D, E in the circuit shown in Figure 5.11.

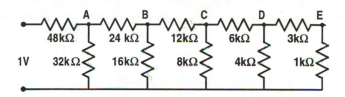

Figure 5.11: Problem 5.3.

5.4 The space between two conductive strips is occupied by a square lattice of $n \times n$ locations as shown in Figure 5.12. Metal balls are placed at random in the lattice. The diameter of the balls is such that they can touch a ball in an adjacent location or the conductive strip. At what percentage occupancy of the lattice will there be a 50% probability of a conductive path being established between the two conductive strips?

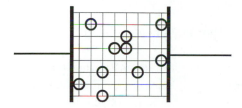

Figure 5.12: Problem 5.4.

Note that the solution to this problem is nontrivial and a computer simulation approach might be more appropriate than an analytic approach. The problem might also be suitable for use as a group exercise. The keyword library index term for further information is percolation theory. (See Chapter 5, *Introduction to Percolation Theory 2nd edn* by Dietrich Stauffer and Amnon Aharony, Taylor & Francis, 1992.)

Unit 6 Power dissipation in resistors

- When a current flows through a circuit, energy is dissipated in the circuit at a rate given by:

$$\text{Power} \quad P = V \times I \quad \text{watts}$$

- For resistive circuits, application of Ohm's law, $V = I \times R$, gives:

$$\text{Power} \quad P = V \times I = I^2 \times R = \frac{V^2}{R} \quad \text{watts}$$

If we go back to the fundamental definition of electrical units, the unit of potential difference is the volt, the unit of charge is the coulomb and the unit of energy is the joule. Consequently, the relationship between the units is that one joule of energy is released or absorbed in moving one coulomb of charge through a potential difference of one volt.

The unit of electrical current is the amp which is equal to a flow of one coulomb of charge per second. The unit of work is the watt and is the rate at which energy is released or generated in joules per second.

The alternatives of release or absorption of energy must be considered because a device such as a battery, a generator or a solar cell drives current around a circuit, doing work on the circuit and releasing energy, whereas a resistor, light bulb or motor will have current driven through it and can be considered to absorb electrical energy from the circuit, the energy then appearing in the form of heat, light or mechanical work done by the motor.

When you select resistors for use in a circuit there are two parameters which you must consider. The first is the resistance, which is marked on the resistor with a colour code system. The second parameter is the maximum power rating for the resistor. This essentially relates to the physical size of the resistor, its ability to withstand heating without damage and its ability to dissipate heat to the surroundings. Most of the resistors used in transistor circuits operate with low voltages across them and with small currents and therefore do not heat up. In most cases the power rating of 0.125 W or one eighth watt is adequate and these resistors are readily available in small sizes. Once the power dissipation rating gets up to 25 W the resistors are usually

24

encased in aluminium housings with bolt holes to permit attaching the resistor to heat sinks to dissipate waste heat. Remember, a typical soldering iron used in electronics is rated at 25 W and this is used to melt the solder holding the circuit together!

6.1 Examples

6.1 A particular torch bulb operates from a 6 V battery and draws a current of 0.5 A. Calculate the total charge which flows around the circuit in 2 minutes operation. Calculate the power rating for the bulb.

$$
\begin{aligned}
\text{Total charge in 2 minutes} &= 0.5\,\text{A} \times 2 \times 60\,\text{s} \\
&= 60\,\text{coulombs} \\
\text{Power dissipation rate in bulb} &= V \times I \quad \text{watts} \\
&= 6\,\text{V} \times 0.5\,\text{A} \\
&= 3.0\,\text{W}
\end{aligned}
$$

6.2 Calculate the current flowing in a 12 V, 60 W car head lamp bulb. Which of the following wire conductor cross sections is used for the connection to the head lamp to prevent any significant heating of the wire: $(0.1\,\text{mm}^2)$, $(0.5\,\text{mm}^2)$, $(1.5\,\text{mm}^2)$, $(5\,\text{mm}^2)$, $(12\,\text{mm}^2)$? Is single strand or multistrand conductor used in this application? Why? (See Problem 3.6.)

$$
\text{Current} = \frac{60\,\text{W}}{12\,\text{V}} = 5\,\text{A}
$$

6.2 Problems

6.1 A particular soldering iron is designed to operate from a 12 V supply and is rated at 25 W. Calculate the current which flows in the element of the soldering iron. Calculate the resistance of the element of the iron when it is at its operating temperature.

6.2 A car head lamp is rated at 80 W and operates on a 12 V supply. Calculate the resistance of the hot filament and the current which flows when the filament is at operating temperature. If the resistance of a tungsten filament increases by a factor of 5 in going from cold to operating temperature, calculate the surge current which flows immediately after switch-on. Explain why bulbs tend to fail at switch-on rather than while they are operating at their normal temperature.

6.3 The circuit in Figure 6.1 shows the resistor chain which is used to obtain the voltages which are applied to the dynodes in a photomultiplier tube. Calculate the voltages at each of the dynodes D_1 to D_7 and also calculate the minimum power rating for the resistors in the circuit. The resistor between the photocathode, PC, and D_1 is $150\,\text{k}\Omega$ and the remainder are all $100\,\text{k}\Omega$.

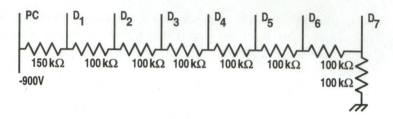

Figure 6.1: Problem 6.3.

6.4 A heat sink is used in electronics to dissipate heat from electronic components and prevent excessive temperature rises. Heat sinks are usually made from extruded aluminium having a large surface area and are rated in degree temperature rise above ambient per watt dissipation. A heat sink rated at $0.5\,^\circ\text{C}$ per watt is used to dissipate the heat from an $18\,\Omega$ resistor in which a current of $0.8\,\text{A}$ is flowing.
Calculate the power dissipated in the resistor.
Calculate the temperature rise above ambient.

6.5 What is the cross section of the copper conductor in the mains lead for a typical piece of portable equipment such as a table lamp? What is the current carrying capacity of the cable? Many mains plugs are rated for a maximum current of $13\,\text{A}$ and, when first purchased, contain $13\,\text{A}$ fuses. Why should the fuse in the plug be changed if $5\,\text{A}$ cable is connected to the plug? What is the function of the fuse?

Unit 7 Power ratios and decibels

- The ratio of the power output from a circuit to the power input to a circuit is quoted in decibels or dB.

$$\text{Power ratio} \quad = 10 \log \left(\frac{P_{out}}{P_{in}} \right) = 20 \log \left(\frac{V_{out}}{V_{in}} \right)$$

The ear and the eye do not have a linear response to sound intensity or brightness. The physiological response is logarithmic and it is found that equal increments in the logarithm of the stimulus give equal increments in sensation. Electrical quantities are therefore often measured on logarithmic scales so as to facilitate calculation of the perceived change of loudness or brightness.

There is also another advantage which is that a very useful compression of scales is obtained when we take the logarithm of the ratio of two quantities. The unit used is the *bel*, named after Alexander Graham Bell. To spread the range somewhat, we can use the *decibel* which is usually abbreviated to dB.

The power dissipated in a circuit is proportional to the square of the voltage. The logarithm of a squared term is 2 times the logarithm of the term; hence the change from 10 to 20 when we go from power ratio to voltage ratio. It is usually assumed that the resistance which comes into the equation $P = \frac{V^2}{R}$ is the same value for the input and the output side but this is not always the case.

In carrying out calculations using the dB scales you must always distinguish carefully between power ratios and voltage ratios and use a multiplier of 10 or 20 as appropriate.

Also, since the dB is essentially a ratio, it is important to specify the reference level so as to facilitate subsequent recovery of the absolute value of the quantity. (See Problem 7.4.)

7.1 Examples

7.1 Calculate the ratio of the output to input for the potentiometer circuit in Figure 7.1.

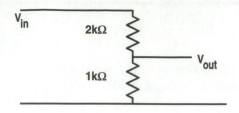

Figure 7.1: Example 7.1.

$$\begin{aligned}
\text{Attenuation} \quad &= \quad 20\log\left(\frac{1\,\text{k}\Omega}{1\,\text{k}\Omega + 2\,\text{k}\Omega}\right) \\
&= \quad -20 \times 0.477 \\
&= \quad -9.54\,\text{dB}
\end{aligned}$$

(Note that a reduction of signal or an attenuation will always give a negative quantity when quoted in dB. We will see later that an amplifier gives a gain which is a positive quantity when quoted in dB.)

7.2 Calculate the value of resistor R in the circuit in Figure 7.2, which will give an attenuation of $-35\,\text{dB}$ between the input and output signals.

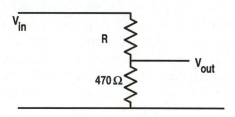

Figure 7.2: Example 7.2.

$$\begin{aligned}
\text{We have} \quad -35 \quad &= \quad 20\log\left(\frac{V_{out}}{V_{in}}\right) \\
&= \quad 20\log\left(\frac{470\,\Omega}{R + 470\,\Omega}\right) \\
\text{Therefore} \quad \frac{470\,\Omega}{R + 470\,\Omega} \quad &= \quad 10^{\left(\frac{-35}{20}\right)} \\
&= \quad 0.01778 \\
\text{giving} \quad 470 \quad &= \quad 470 \times 0.01778 + R \times 0.01778 \\
\text{and} \quad R \quad &= \quad 25.96\,\text{k}\Omega
\end{aligned}$$

7.2 Problems

7.1 Calculate the attenuation in dB of the potential divider in Figure 7.3.

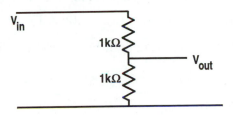

Figure 7.3: Problem 7.1.

7.2 Calculate the values for the resistors in a potential divider which has an attenuation of $-50\,\text{dB}$ and where the total resistance in the divider is $25\,\text{k}\Omega$.

7.3 Calculate the output voltage from the circuit in Figure 7.4 for $V_{in} = 3\,\text{V}$. Calculate the attenuation of this circuit. (Suggested approach: Find the voltage at point X and then use this voltage to find the voltage at the output.)

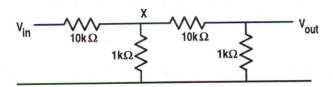

Figure 7.4: Problem 7.3.

7.4 The threshold of hearing is, by convention, taken to be at a sound intensity of $10^{-12}\,\text{Wm}^{-2}$ and this is assigned the level on the dB scale of $0\,\text{dB}$. Calculate the sound intensity for a conversation for which the sound intensity level is $60\,\text{dB}$ and for a position near the end of a runway with an aircraft taking off for which the sound intensity level is $120\,\text{dB}$.

Unit 8   AC and DC waveforms

Three numbers are used to specify a sinusoidal waveform:

- Amplitude of the waveform, either voltage, V_0, or current, I_0

- Frequency of the waveform, f, in hertz (Hz) or angular frequency, ω, (radians per second)

  OR

  The period, T, in seconds (s) given by $T = \frac{1}{f}$

- The phase, ϕ, in degrees or radians, measured with respect to a reference waveform of the same frequency:

$$\omega = 2\pi f$$
$$T = \frac{1}{f}$$
$$V = V_0 \sin(2\pi ft + \phi) \quad \text{for voltage waveforms}$$
$$I = I_0 \sin(2\pi ft + \phi) \quad \text{for current waveforms}$$

Calculations are usually made using radian mode.

The sign of the phase shift is the same as the sign of the output waveform measured at the time of a positive going zero crossing of the reference sine wave. The phase angle of the lower waveform in Figure 8.1 is negative.

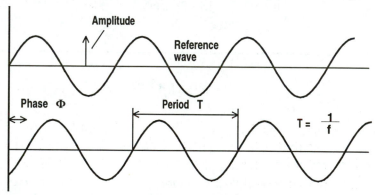

Figure 8.1: Specification of sinusoidal waveforms.

The single cells or batteries of cells in series which we have used in the circuit diagrams give constant output voltage resulting from the conversion of chemical energy into electrical energy. This is a very inefficient process and would not be capable of supplying the electrical energy used in a modern society.

The conversion of mechanical energy to electrical energy in a generator is a much more efficient process and also does not leave a residue of used batteries to be disposed of! The operation of a generator depends on Lenz's law which states that when the magnetic flux through a conducting loop changes, a voltage is generated or induced in the conductor which is proportional to the rate of change of the magnetic field, B, the area of the loop, A, and the number of turns of conductor, N. The induced emf, $\mathcal{E}$, is then given by:

$$\mathcal{E} = -NA\frac{dB}{dt}$$

The sign of the induced emf is such that it would drive a current in the conducting loop which would oppose the change in magnetic flux intersecting the loop. The work done by the mechanical motor or hydroelectric source of power in moving the loop in the magnetic field is converted into electrical energy at the output of the generator. The rotary motion of the conducting loop causes the magnetic flux direction through the loop to alternate which gives an alternating voltage waveform at the output of the generator and a consequent alternating current. In electronics we will often use alternating voltages which do not originate from rotation of coils in magnetic fields but the description of the waveforms is the same.

The power companies in Europe supply electrical current at a frequency of 50 Hz. In America the frequency is 60 Hz. In some specialist applications such as aircraft power systems the line frequency can be higher, 400 Hz, since this reduces the weight of transformers and generators. On some railway systems DC or low frequency AC is sometimes used since DC or low frequency AC electrical motors give greater torques at low speeds than can be obtained from normal AC induction motors. In electronics the frequencies used extend over a much wider range, loosely described as audio for 1 kHz, RF (radio frequency) for 100 MHz and microwave for 10,000 MHz.

The one great advantage of using AC is that it is possible to change the voltage either up or down by the use of a transformer without any significant loss of power. Voltage conversion in DC is much more difficult, inefficient and expensive in terms of the equipment required (motor-generators or inverters).

The term, AC, stands for alternating current and therefore the use of the term AC as a descriptor for a voltage is not a strictly valid usage but you will encounter the term AC voltages instead of the more correct term, alternating voltages.

Most electronic circuits are analyzed in terms of their response to sinusoidal waveforms. However, there are many situations where other waveforms are used. The most common are square, triangle and sawtooth waveforms, as shown in Figure 8.2. The mathematical description of these waveforms is not as simple as the sinusoidal waveform and the response of circuits to these waveforms is not as susceptible to mathematical analysis.

In specifying the phase, ϕ, of a sine wave, a reference sine wave of the same frequency is used. When this reference wave crosses zero in the positive going direction, for example at the origin in the diagram in Figure 8.1, the value of the sign of the second wave is observed. If the sign is positive then the phase, ϕ, is positive; if the sign is negative, as it is in Figure 8.1, then the phase is negative. If the second wave has a positive value of phase, it is said to **lead** the reference; if the second wave has a negative value of phase, it is said to **lag** the reference wave. Therefore, if the magnitude of the phase shift in Figure 8.1 is 60° or 1.05 radians, the equation which describes the second wave in Figure 8.1 is:

$$Y = Y_0 \sin(2\pi f t - 1.05)$$

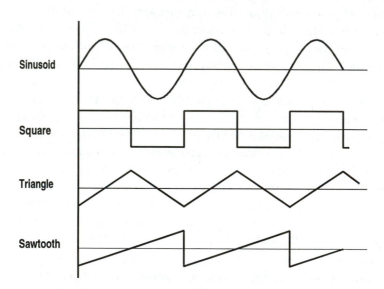

Figure 8.2: Common waveforms.

8.1 Example

8.1 Figure 8.3 shows a sketch of a sinusoidal voltage waveform displayed on an oscilloscope which has a Y axis setting of 2 V/division and a time axis setting of 5 ms/division. Calculate the amplitude, frequency, angular frequency and period of the waveform and write down the equation for the voltage waveform.

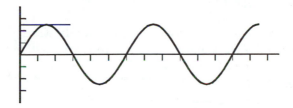

Figure 8.3: Example 8.1.

The maximum of the waveform is at 2.5 divisions and therefore the amplitude of the waveform is $2.5 \times 2\,\mathrm{V} = 5.0\,\mathrm{V}$.

The period of the waveform is 6 divisions and therefore $T = 6 \times 5 \times 10^{-3}$ seconds $= 30\,\mathrm{ms}$.

The frequency is $f = \frac{1}{T} = \frac{1}{30 \times 10^{-3}}\,\mathrm{Hz} = 33.3\,\mathrm{Hz}$.

The angular frequency is $\omega = 2\pi f = 209.3$ radians per second.

There is no reference waveform so we take the phase $\phi = 0$.

The equation which describes the waveform is therefore:

$$V = 5.0\sin(209.3t + 0) = 5.0\sin(209.3t) \text{ volts}$$

8.2 Problems

8.1 A particular sinusoidal voltage waveform has parameters as follows:

$$
\begin{aligned}
\text{Voltage amplitude} \quad V_0 &= 13\,\text{volts}\\
\text{Frequency} \quad f &= 1.5\,\text{kHz}\\
\text{Phase} \quad \phi &= +0.6\,\text{radians}
\end{aligned}
$$

Give a scaled sketch of a reference voltage waveform and of this voltage waveform for time from $t = 0$ to $t = 10\,\mathrm{ms}$. Calculate the times when the voltage is zero. Calculate the period of the waveform.

8.2 The magnetic field, B, between the pole pieces of a permanent magnet is 0.15 tesla. A circular search coil of diameter 12 mm and containing 200 turns of wire is placed between the poles. Calculate the average voltage induced in the coil when the coil is withdrawn from between the poles in a time of 0.1 seconds.

8.3 Convert $47°$ to radians.

Convert 2.45 rad to degrees.

8.4 A waveform is specified as having a periodic time, $T = 2.6$ ms. The voltage is $+5$ V for times from $n \times T$ to times $(n + \frac{1}{2})T$ and the voltage is -5 V for times from $(n + \frac{1}{2}) \times T$ to times $(n + 1) \times T$ where $n = 0, 1, 2, 3, \ldots$ Give a scaled sketch of the voltage waveform.

8.5 A particular sinusoidal voltage waveform has an amplitude of $V_0 = 7.3$ V, a frequency of 170 Hz and a phase of $\phi = 0.4$ rad.
Calculate the values of the voltage at times 0 s, 14 ms, 2.4 s and 10.2 s.

8.6 A triangular voltage waveform has an amplitude of 4.3 V, a phase angle of -0.1 rad measured with respect to a reference waveform and a frequency of 280 Hz.
Sketch the waveform and calculate the values of the voltages at times 2.0 ms, 10.2 ms and 180.0 ms.

8.7 A square voltage waveform has an amplitude of 4 V, a periodic time of 120 ms and crosses the zero volts level in the positive going direction at time $t = 0.1$ s.
Sketch the voltage waveform and calculate the value at $t = 490$ ms.

Unit 9   Voltage and power in AC circuits

- When the magnitude of the voltage waveform is specified in volts RMS (root mean square), the average power dissipated in a resistor, R, is given by:

$$\text{Power} \quad P = \frac{V_{RMS}^2}{R} = I_{RMS}^2 \times R \quad \text{watts}$$

- For a sinusoidal waveform:

$$V_{Amplitude} = 1.4 \times V_{RMS}$$
$$V_{Peak-to-Peak} = 2 \times V_{Amplitude}$$

At any instant, the power dissipation in an electrical component is given by $P = V \times I$. For the special case of a resistor, we have Ohm's law, $V = I \times R$, and when we substitute for I, we get $P = \frac{V^2}{R}$. However, we also need to know the average power dissipation. This can be calculated, in the case of a sinusoidal waveform, by integrating the instantaneous power dissipation over a cycle and then averaging over the periodic time, T.

$$
\begin{aligned}
P_{Average} &= \frac{1}{T} \int_0^T \frac{V^2}{R} dt \\
&= \frac{1}{T} \int_0^T \frac{V_0^2 \sin^2(\frac{2\pi t}{T})}{R} dt \\
&= \frac{1}{T} \int_0^T V_0^2 \frac{1 - \cos\frac{4\pi t}{T}}{2R} dt \\
&= \frac{1}{T} \left[\frac{V_0^2 t}{2R} + \frac{V_0^2 T \sin\frac{4\pi t}{T}}{8\pi R} \right]_0^T \\
&= \frac{V_0^2}{2R}
\end{aligned}
$$

$$\text{Let} \quad 1.414 \times V_{RMS} = V_0$$

$$\text{which gives} \quad P_{Average} = \frac{V_{RMS}^2}{R}$$

$$\text{Form factor for sinusoids} = 1.414 = \sqrt{2}$$

When a meter is used to measure alternating voltage, the value indicated is usually in volts root mean square. When an oscilloscope is used to measure voltages, the most convenient operation is to measure the peak-to-peak voltage and use the form factor to convert to volts rms. It is important to specify the type of measurement and the type of voltage unit in use. If a nonsinusoidal voltage waveform is present then the integration shown above will be different and the form factor used in the conversion will be different from the 1.4 used with sinusoidal waveforms.

Some of the digital storage oscilloscopes on the market have facilities which allow you to display a waveform and also have an indication at the bottom of the screen which gives the numerical value of voltage magnitude in either *pp* or *rms* according to a selection on a menu. If you get an opportunity, you should try out this feature on a digital oscilloscope.

9.1 Example

9.1 Calculate the RMS voltage for the oscilloscope tracing of the voltage waveform shown in Figure 9.1. Calculate the frequency. The oscilloscope settings are $2\,\text{ms/division}$ for the time base and $5\,\text{V/division}$ for the Y amplifier.

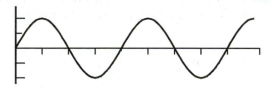

Figure 9.1: Example 9.1.

From the figure the amplitude is $2 \times 5\,\text{V}$ and the period is $4 \times 2 \times 10^{-3}\,\text{s}$.

$$V_{RMS} = \frac{2 \times 5}{1.41} = 7.07\,\text{V}$$

$$\text{and} \quad \text{Frequency} = \frac{1}{\text{Period}} = \frac{1}{4 \times 2 \times 10^{-3}}$$

$$= 125\,\text{Hz}$$

9.2 Problems

9.1 Calculate the RMS voltage for a $6\,\mathrm{V_{pp}}$, $1\,\mathrm{kHz}$ sinusoidal waveform. Will the RMS Voltage change if the frequency changes to $1.5\,\mathrm{kHz}$?

9.2 Give a scaled sketch of the waveform which you would observe on an oscilloscope which is displaying an $11\,\mathrm{V_{RMS}}$, $300\,\mathrm{Hz}$ sinusoidal waveform.

9.3 Plot a graph of a $400\,\mathrm{kHz}$, $6\,\mathrm{V}$ amplitude square wave. Use either a square counting method or integration to calculate the RMS voltage. What is the form factor for a square wave?

9.4 A $500\,\mathrm{Hz}$ square waveform, of amplitude $8\,\mathrm{V}$, is superimposed on a $9\,\mathrm{V}$ DC voltage. Calculate the RMS voltage. Is the RMS voltage greater when there is a DC component present in the signal?

9.5 Calculate the form factor for the trapezoidal voltage waveform shown in Figure 9.2.

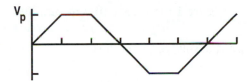

Figure 9.2: Problem 9.5.

Unit 10   Capacitors

- The voltage, in volts, across a capacitor, in farads, is related to the charge, in coulombs, by:
$$Q = C \times V$$

- When capacitors are connected in parallel the resultant capacitance is given by:
$$C_p = C_1 + C_2 + C_3 + \cdots$$

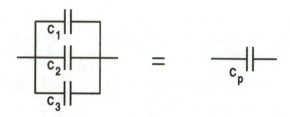

Figure 10.1: Capacitors in parallel.

- When a sinusoidal voltage of the form:
$$V = V_0 \sin(2\pi f t)$$

is applied across a capacitor, the current is 90° or $\frac{\pi}{2}$ radians out of phase with the driving voltage and is given by:
$$I = CV_0 2\pi f \sin(2\pi f t + \frac{\pi}{2})$$

The fundamental defining equation for a capacitance, $Q = CV$, can be differentiated to obtain the current:
$$I = \frac{dQ}{dt} = C\frac{dV}{dt}$$

Therefore the current through a capacitor is proportional to the capacitance and to the rate of change of the voltage across the capacitor. But you should note that the electrons that enter one lead of the capacitor are not the same

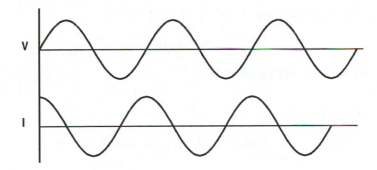

Figure 10.2: Steady state voltage and current waveforms in a capacitor.

electrons that emerge from the other lead. There is, however, a net transfer of charge through the capacitor.

Suppose we apply a sinusoidal voltage waveform across a capacitor.

$$V = V_0 \sin(2\pi f t)$$
$$\text{We then get} \quad I = C\frac{dV}{dt}$$
$$= CV_0 2\pi f \cos(2\pi f t)$$
$$= CV_0 2\pi f \sin(2\pi f t + \frac{\pi}{2})$$

Therefore, for a sinusoidal waveform, the voltage across a capacitor and the current through a capacitor are 90° or $\frac{\pi}{2}$ out of phase with each other. From the definition of phase angle in Unit 8 we see that the phase angle of the current waveform is positive with respect to the voltage waveform which we take as the reference and the current waveform therefore leads the voltage waveform.

In Unit 6, we saw that the power dissipation in a resistor is given by $P = V \times I$. We use the same equation to calculate the power dissipation in a capacitor. We have, for the instantaneous power dissipation:

$$P = V \times I$$
$$= V_0 \sin(2\pi f t) \times CV_0 2\pi f \sin(2\pi f t + \frac{\pi}{2})$$
$$= CV_0^2 2\pi f \sin(2\pi f t) \sin(2\pi f t + \frac{\pi}{2})$$
$$= CV_0^2 2\pi f \sin(2\pi f t) \cos(2\pi f t)$$
$$= CV_0^2 \pi f \sin(4\pi f t)$$

The average power is then obtained by integration:

$$P_{Ave} = \frac{1}{T} \int_0^T CV_0^2 \pi f \sin(4\pi f t) dt = 0$$

so that no power is dissipated in a capacitor.

10.1 Example

10.1 If a $15\,V_{Amplitude}$, $20\,kHz$ sinusoidal voltage is applied across a $0.1\,\mu F$ capacitor, calculate the current in the capacitor.

The current is calculated as follows:

$$
\begin{aligned}
I &= CV_0 2\pi f \sin(2\pi f t + \frac{\pi}{2}) \\
&= 0.1 \times 10^{-6} \times 15 \times 2\pi \times 20 \times 10^3 \times \sin(2\pi \times 20 \times 10^3 \times t + \frac{\pi}{2}) \\
&= 0.188 \sin(1.256 \times 10^5 \times t + 1.57)\,A_{Amplitude}
\end{aligned}
$$

and the expected oscilloscope traces of the input voltage and output current waveforms are shown in Figure 10.3.

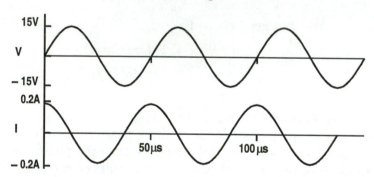

Figure 10.3: Example 10.1.

10.2 Problems

10.1 A $100\,\mu F$ capacitor is connected across the $220\,V_{RMS}$, $50\,Hz$ mains supply. Calculate the RMS current which flows in the capacitor. (Such capacitors are sometimes used for power factor correction and help to keep the voltage and current in phase when large motors are in use.)

10.2 A voltage of $3\,V_{pp}$ at $10\,MHz$ is applied across a $2\,nF$ capacitor. Calculate the current in the capacitor. If the frequency is increased to $15\,MHz$, calculate the new current.

Unit 11   Inductors

- The unit of inductance is the henry and the circuit symbol used for the inductor is:

Figure 11.1: Circuit symbol for inductor.

- A rate of change of current of one amp per second through an inductor of one henry gives a voltage across the inductor of one volt.

- If the current which flows through an inductor is given by:

$$I = I_0 \sin(2\pi f t)$$

then the voltage across the inductor is:

$$V = L 2\pi f I_0 \sin(2\pi f t + \frac{\pi}{2})$$

An inductor is basically a coil of wire, with or without an iron or ferrite core. The current through the wire sets up a magnetic field for which the flux lines loop through the coil. Energy is stored in this magnetic field. If the current in the coil is changed then energy is either added to or removed from the energy stored in the magnetic field. Therefore in carrying out calculations on inductors, we are concerned with the changes in the current through the inductor.

If a current is set flowing in a coil made from superconducting metal (very low temperatures) then the current will flow indefinitely and the magnetic field of the coil will remain constant. This feature is used in the superconducting magnets, operated at liquid helium temperatures, which are found in the magnetic resonance tomography scanners used in hospitals.

At room temperatures, the resistance of the wires will usually cause the current to die away in a short time and the energy stored in the magnetic field will be dissipated in resistive heating of the coil.

The laws of electromagnetic induction, which derive from Lenz's law, give the relationship between the rate of change of current in a coil and the induced emf:

$$\mathcal{E} = -L\frac{dI}{dt}$$

where the unit of inductance, H, is the henry.

If an external voltage is applied across an inductor so as to give a current in the inductor which varies sinusoidally in accordance with:

$$I = I_0 \sin(2\pi ft)$$

then from Lenz's law, this externally applied voltage is opposed by the induced emf:

$$
\begin{aligned}
V &= -\mathcal{E} \\
&= L\frac{dI}{dt} \\
&= L\frac{d}{dt}\left(I_0 \sin(2\pi ft)\right) \\
&= L2\pi f I_0 \cos(2\pi ft) \\
&= L2\pi f I_0 \sin(2\pi ft + \frac{\pi}{2}) \\
&= L\omega I_0 \sin(\omega t + \frac{\pi}{2})
\end{aligned}
$$

$$\text{where} \quad \omega = 2\pi f \quad \text{is the angular frequency}$$

Therefore, for a sinusoidal current through an inductor, the voltage across the inductor is 90° out of phase with the current waveform.

There is one significant difference between inductors and resistors or capacitors. It is possible to combine inductors in series or parallel and calculate a single equivalent inductor only when the inductors are sufficiently separated so that no inductor is in the magnetic field of another inductor.

There are two main types of inductor:

- **Air cored inductors** where the wire used is stiff enough to support itself or else is wound on a plastic or paper supporting former. For air cored inductors, the inductance in nH $= 10^{-9}$ H is given by:

$$L(\text{nH}) = \frac{N^2 d^2}{0.46d + 1.02b}$$

where N is the number of turns, d is the coil diameter in mm and b is the length of the coil in mm.

- **Ferrite cored inductors** have higher values for the inductance because of the higher magnetic permeability of the ferrite core. The ferrite core manufacturer will usually give a formula for the inductance which takes into account the ferrite type, size etc. A typical formula would be:

$$L(\text{nH}) = N^2 A_L$$

where L is the inductance in nH, N is the number of turns of wire in the coil and A_L is the inductance factor which is provided by the manufacturer of the ferrite core. Ferrite cores usually have an upper operating frequency due to the limited frequency response of the ceramic ferrite material.

11.1   Example

11.1 A sinusoidally varying current of $2\,\text{mA}$ amplitude and frequency $3\,\text{kHz}$ passes through a coil having an inductance of $10\,\text{mH}$. Calculate the voltage across the inductor and graph the current and voltage waveforms.

$$\begin{aligned} V &= L\frac{dI}{dt} = L\frac{d}{dt}(I_0 \sin(2\pi ft)) = L2\pi f I_0 \sin(2\pi ft + \frac{\pi}{2}) \\ &= 0.01 \times 2\pi 3 \times 10^3 \times 2 \times 10^{-3} \times \sin(2\pi 3 \times 10^3 \times t + \frac{\pi}{2}) \\ &= 0.377 \sin(18849t + 1.57) = 0.377 \cos(18849t) \end{aligned}$$

The current and voltage waveforms are shown in Figure 11.2.

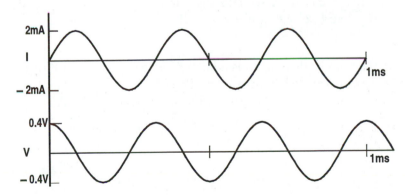

Figure 11.2: Example 11.1. Current and voltage waveforms for inductor.

Compare these current and voltage waveforms for an inductor shown in Figure 11.2 to the voltage and current waveforms for a capacitor shown in Figure 10.2 and you will see that the phase shifts of the current waveforms are of opposite sign for capacitors and inductors.

11.2   Problems

11.1 If the current in a 0.35 H inductor changes at a steady rate of $0.8\,\mathrm{A\,s^{-1}}$, calculate the voltage across the inductor. Assume that the inductor has zero resistance.

11.2 A voltage of 20 V is applied across a 0.015 H inductor. Calculate the initial rate of change of current in the inductor.

11.3 Calculate the inductance of an air cored inductor having 30 turns, a diameter of 12 mm and a length of 20 mm.

11.4 An inductor of value 2.9 μH is required for a radio frequency tuning circuit. If the inductor is to be wound from 0.5 mm diameter wire which is stiff enough to give a self supporting coil, calculate suitable dimensions for the coil and the number of turns required. Is there a single, unique answer?

11.5 Calculate the number of turns required for an inductor having a value of 590 μH constructed using a ferrite core type for which the manufacturer quotes an inductance factor, $A_L = 250$.

11.6 A sinusoidal voltage of 6 V amplitude at 1 kHz is applied across a 200 mH inductor. Calculate the current in the inductor and plot the current and voltage waveforms. Obtain the expression for the current waveform.

11.7 If the amplitude of a sinusoidal current waveform in a 30 mH inductor is 12 mA and the frequency is 2 kHz, calculate the amplitude of the voltage across the inductor. Plot the current and voltage waveforms and show the phase difference. Obtain the equations for the current and for the voltage.

Unit 12 Complex impedance of R, C and L

- The general form of Ohm's law is:

$$V = Z \times I$$

  where Z is the complex impedance.

  - The impedance of a resistor is R.

  - The impedance of a capacitor is $\frac{1}{j\omega C}$.

  - The impedance of an inductor is $j\omega L$.

- The units of impedance are ohms.

In the last two units we have seen that for a capacitor and for an inductor the current and voltage are out of phase by 90° or $\frac{\pi}{2}$. For a resistor, the voltage and the current are in phase.

The representation of the waveforms by trigonometric functions, such as sin and cos, is cumbersome and a more elegant approach is to use complex numbers, where we use the relationship:

$$e^{j\theta} = \cos\theta + j\sin\theta$$

where $j = \sqrt{-1}$. (Note that in electronics we use j rather than i to represent $\sqrt{-1}$ because of the possibility of confusion with i when it is used to represent a current.)

A sinusoidally varying voltage can then be represented by the imaginary part of:

$$
\begin{aligned}
V &= V_0 e^{j\omega t} \\
\text{and then} \quad \frac{dV}{dt} &= V_0 j\omega e^{j\omega t} \\
&= j\omega V
\end{aligned}
$$

$$\text{So for capacitors} \quad I = C\frac{dV}{dt}$$

$$= j\omega CV$$

$$\text{giving} \quad V = \frac{1}{j\omega C}I$$

$$\text{And for inductances} \quad L\frac{dI}{dt} = V$$

$$\text{By integration this becomes} \quad LI = \int V\,dt$$

$$= \frac{1}{j\omega}V$$

$$\text{Giving} \quad V = j\omega LI$$

We can now generalize Ohm's law to get:

$$V = ZI$$

$$\text{where complex impedance} = Z$$

$$\text{and} \quad Z_R = R \quad \text{for a resistance}$$

$$Z_C = \frac{1}{j\omega C} \quad \text{for a capacitor}$$

$$\text{and} \quad Z_L = j\omega L \quad \text{for an inductor}$$

These three results will be used throughout the text.

12.1   Example

12.1 Calculate the impedance of a $0.1\,\mu$F capacitor at a frequency of $19\,$kHz.

The impedance of a capacitor is:

$$Z_C = \frac{1}{j\omega C}$$

$$= \frac{1}{j2\pi f C}$$

$$= \frac{-j}{2\pi f C}$$

$$= \frac{-j}{2\pi 19 \times 10^3 \times 0.1 \times 10^{-6}}$$

$$= \frac{-j}{0.01193}$$

$$= -83.8j\,\Omega$$

In this result, the 83.8 gives the numerical relationship between the magnitude or amplitude of the voltage and current waveforms. The

j indicates that the voltage and current sinusoidal waveforms are 90° out of phase with each other and the − sign indicates that the current waveform leads the voltage waveform. (Review Unit 8 and also examine Figure 10.2.)

12.2 Problems

12.1 Calculate the complex impedance of a 100 nF capacitor at frequencies of 1 kHz and 80 kHz.

12.2 Calculate the complex impedance of a 5 mH inductor at 200 Hz and at 100 kHz.

12.3 As the frequency increases, does the magnitude of the impedance of a capacitor increase or decrease?

12.4 As the frequency increases, does the magnitude of the impedance of an inductor increase or decrease?

12.5 A sinusoidal voltage waveform of amplitude 3.1 V and frequency 22.9 kHz is applied across a 0.22 μF capacitor.
Write down the equation for the voltage waveform in complex notation. Calculate the complex impedance of the capacitor. Calculate the equation for the current in the capacitor expressed in complex notation.

12.6 A sinusoidal voltage waveform is applied across a capacitor. Calculate the average power dissipated in the capacitor by integrating the product of the voltage across the capacitor and the current through the capacitor.

12.7 A sinusoidal voltage waveform is applied across an inductor. Calculate the average power dissipated in the inductor by integrating the product of the voltage across the inductor and the current through the inductor.

Unit 13 Complex impedance diagram

- The impedance of components can be represented on a complex impedance diagram by vectors drawn from the origin.

- When components are connected in series, the resultant impedance is obtained from the vector sum of the impedances of the individual components.

In drawing the complex impedance diagram the following rules are used:

- A resistance is represented by a vector of magnitude R ohms drawn along the positive x axis.

- A capacitor is represented by a vector of magnitude $\frac{1}{\omega C}$ ohms drawn along the negative y axis.

- An inductance is represented by a vector of magnitude ωL ohms drawn along the positive y axis.

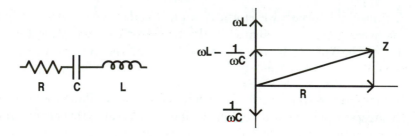

Figure 13.1: Complex impedance diagram for RCL in series.

The resultant impedance, Z, of the three components connected in series is the vector sum as shown in the diagram.

The magnitude of the impedance of the resultant is obtained from:

$$|Z| = \sqrt{R^2 + \left(\omega L - \frac{1}{\omega C}\right)^2}$$

48

The phase angle, ϕ, is obtained from:

$$\phi = \tan^{-1}\left(\frac{\omega L - \frac{1}{\omega C}}{R}\right)$$

So we now have the result that the voltage and the current are related by:

$$\begin{aligned} V &= ZI \\ &= |Z|\,e^{j\phi}I \\ \text{or}\quad V_0 e^{j\omega t} &= |Z|\,I_0 e^{j(\omega t + \phi)} \end{aligned}$$

The sign of the phase angle can be determined by using the equation for ϕ given above or by using the diagram shown in Figure 13.2.

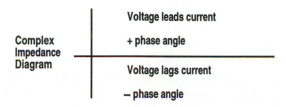

Complex Impedance Diagram

Voltage leads current
+ phase angle

Voltage lags current
− phase angle

Figure 13.2: Identification of the sign of the phase angle.

In this diagram, the current waveform is taken as the reference and the sign of the phase angle is + when the resultant impedance lies in the top half of the diagram and − when the resultant impedance lies in the bottom half of the diagram. This is probably best remembered by saying that a current through a component gives a voltage across the component.

If $I = I_0 \sin(2\pi f t)$ then $V = V_0 \sin(2\pi f t + \phi)$.

13.1 Example

13.1 Calculate the complex impedance of 100 nF in series with 820 Ω at a frequency of 3.5 kHz.

The magnitude of the impedance is given by:

$$\begin{aligned} |Z| &= \sqrt{820^2 + \left(\frac{1}{2\pi 3.5 \times 10^3 \times 100 \times 10^{-9}}\right)^2} \\ &= 938\,\Omega \end{aligned}$$

and the phase angle is given by:

$$\phi = \tan^{-1}\left(\frac{\frac{-1}{2\pi f C}}{R}\right)$$

$$= \tan^{-1}\left(\frac{-1}{2\pi f C R}\right)$$

$$= -0.51 \text{ radians} \quad \text{or} \quad -29°$$

The voltage, V, and the current, I, will then be related by the equation:

$$V = 938 \times e^{-0.51j} I$$

Note that when exponential notation is used the value for ϕ is always in radians.

13.2   Problems

13.1 Verify that the waveform diagrams shown in Figures 10.2 and 11.2 give signs of the phase angle which are in agreement with the signs of the phase angles obtained from Figure 13.2.

13.2 Plot on a complex impedance diagram the impedances of a $2.2\,\text{k}\Omega$ resistor and a $0.2\,\mu\text{F}$ capacitor for a frequency of $1.2\,\text{kHz}$. Use the plot to estimate the magnitude of the impedance and the phase angle when these two components are connected in series.

13.3 Plot the complex impedance diagram for $3.3\,\text{k}\Omega$ in series with $47\,\text{mH}$ at a frequency of $1.3\,\text{kHz}$. Calculate the magnitude of the impedance and the phase angle.

13.4 Plot the locus or path of the tip of the impedance vector for a resistor of $1.5\,\text{k}\Omega$ in series with a capacitance of $0.1\,\mu\text{F}$ as the frequency is varied from $100\,\text{Hz}$ to $50\,\text{kHz}$.

13.5 An inductance of $0.1\,\text{mH}$, a resistance of $680\,\Omega$ and a capacitance of $0.22\,\mu\text{F}$ are connected in series. Sketch the path followed on the complex impedance diagram by the tip of the resulting complex impedance vector as the frequency is varied from $10\,\text{kHz}$ to $90\,\text{kHz}$.

Unit 14 Resistances and reactances

- Any combination of resistors, capacitors and inductors, when driven by a sinusoidal signal at an angular frequency ω, can be combined to give a resultant impedance in the form:

$$Z = R + jX$$

where R is the resistance and X is the reactance.

When resistors are connected in series, the resultant resistance is the sum of the individual resistances:

$$R_S = R_1 + R_2 + R_3 + \cdots$$

When resistors, capacitors and inductors are connected in series the resultant impedance is the sum of the individual impedances:

$$Z_S = R_1 + R_2 + \cdots + \frac{1}{j\omega C_1} + \frac{1}{j\omega C_2} + \cdots + j\omega L_1 + j\omega L_2 + \cdots$$

This can be reduced, using complex algebra, to give the impedance, Z, as the sum of a resistive component, R, and a reactive component, X.

$$Z_S = R_S + jX_S$$

The units of Z, R and X are ohms.

Similarly, when components are connected in parallel the equivalent of:

$$\frac{1}{R_P} = \frac{1}{R_1} + \frac{1}{R_2} + \cdots$$

applies with the R_n replaced by the component complex impedance.

Note that the impedance of a capacitor is $\frac{1}{j\omega C}$ and the impedance of an inductor is $j\omega L$. Both of these terms depend on the frequency of the voltage across the component and therefore the impedance changes if the frequency changes. Also the waveform is presumed to be of sinusoidal form and to have been applied for long enough for any start-up transients to have died away and a steady state reached.

14.1   Examples

14.1 Calculate the impedance of a $0.1\,\mu\mathrm{F}$ capacitor connected in series with an $820\,\Omega$ resistance at a frequency of $1\,\mathrm{kHz}$.

$$
\begin{aligned}
Z \;&=\; R + \frac{1}{j\omega C} \\[4pt]
&=\; R + \frac{1}{j2\pi f C} \\[4pt]
&=\; 820 + \frac{1}{j2\pi 1000 \times 0.1 \times 10^{-6}} \\[4pt]
&=\; 820 - \frac{j}{0.000628} \\[4pt]
&=\; 820 - j1592
\end{aligned}
$$

$$
\begin{aligned}
\text{Resistance}\quad R &= 820\,\Omega \\
\text{Reactance}\quad X &= -1592\,\Omega \\
\text{Impedance}\quad Z &= 820\,\Omega - j1592\,\Omega
\end{aligned}
$$

Verify the change of sign in the fourth line.

14.2 Calculate the impedance of a $1\,\mathrm{M}\Omega$ resistor in parallel with a $30\,\mathrm{pF}$ capacitor at $40\,\mathrm{kHz}$. (This is the equivalent input impedance for an oscilloscope.)

$$
\begin{aligned}
\frac{1}{Z} \;&=\; \frac{1}{R} + j\omega C \\[4pt]
&=\; \frac{1}{10^6} + j2\pi 40 \times 10^3 \times 30 \times 10^{-12} \\[4pt]
&=\; 10^{-6} + j7.54 \times 10^{-6}
\end{aligned}
$$

$$
\begin{aligned}
\text{Therefore}\quad Z \;&=\; \frac{1}{10^{-6} + j7.54 \times 10^{-6}} \\[4pt]
&=\; \frac{10^{-6} - j7.54 \times 10^{-6}}{(10^{-6})^2 + (7.54 \times 10^{-6})^2} \\[4pt]
&=\; \frac{10^{-6} - j7.54 \times 10^{-6}}{5.78 \times 10^{-11}} \\[4pt]
&=\; 17\,k\Omega - j130\,k\Omega
\end{aligned}
$$

14.2 Problems

14.1 Calculate the impedance of $320\,\Omega$ in series with $10\,\text{mH}$ at $30\,\text{kHz}$.

14.2 Calculate the impedance of $0.2\,\mu\text{F}$, $6\,\text{mH}$ and $680\,\Omega$ all connected in series at a frequency of $4\,\text{kHz}$.

14.3 Calculate the impedance of $2.2\,\text{k}\Omega$ connected in parallel with $0.1\,\mu\text{F}$ at a frequency of $2\,\text{kHz}$.

14.4 Calculate the impedance of a resistance of $12\,\Omega$ connected in parallel with an inductance of $0.2\,\text{H}$ at a frequency of $50\,\text{Hz}$.

14.5 Calculate the impedance of $680\,\Omega$, $0.5\,\mu\text{F}$ and $10\,\text{mH}$ all connected in parallel at a frequency of $7\,\text{kHz}$.

14.6 Convert the complex impedance from the form:

$$Z = R + jX$$

to the form:

$$Z = |Z|\,e^{j\phi}$$

14.7 A sinusoidal voltage waveform is applied across a complex impedance, $Z = R + jX$. Obtain an expression for the average power dissipated in the complex impedance, expressed in terms of the resistance and reactance.

Unit 15 Generalized potential divider

- Resistive or reactive components used in potential dividers give:

$$V_{out} = \frac{Z_2}{Z_1 + Z_2} V_{in}$$

- When the term $\frac{Z_2}{Z_1+Z_2}$ is put into the form $|A|\, e^{j\phi}$ then $|A|$ is the attenuation of the potential divider and ϕ is the phase shift.

There are two results from complex algebra which we will use extensively and which you may need to follow up in your mathematics textbook.

When a complex number, c, is in the form $c = a + jb$, the modulus and the phase angle for c and $\frac{1}{c}$ are given by:

$$|c| = |a + jb| = \sqrt{a^2 + b^2} \quad \text{and} \quad \tan\phi = \frac{b}{a}$$

$$\left|\frac{1}{c}\right| = \left|\frac{1}{a + jb}\right| = \frac{1}{\sqrt{a^2 + b^2}} \quad \text{and} \quad \tan\phi = \frac{-b}{a}$$

The resistors in the potential divider discussed in Unit 4 can be replaced by any combination of resistors, capacitors or inductors in series or parallel. A resultant impedance can then be calculated for each half of the potential divider. The current in each of the two impedances is given by $\frac{V_{in}}{Z_1+Z_2}$. This current flowing through Z_2 gives an output voltage $Z_2 I$.

The ratio of output to input voltage is then:

$$\frac{V_{out}}{V_{in}} = \frac{Z_2}{Z_1 + Z_2}$$

but since Z_1 and Z_2 are complex then $\frac{Z_2}{Z_1+Z_2}$ is usually also complex and has a magnitude less than 1.

If we express $\frac{Z_2}{Z_1+Z_2}$ in the form $|A|\, e^{j\phi}$ then $|A|$ gives the attenuation of the potential divider and ϕ gives the phase shift in radians.

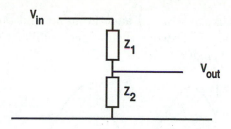

Figure 15.1: Generalized potential divider.

15.1 Example

15.1 Calculate the attenuation and phase shift in the RC network in Figure 15.2 where $R = 2.2\,\text{k}\Omega$, $C = 0.1\,\mu\text{F}$ and the frequency is $1.5\,\text{kHz}$.

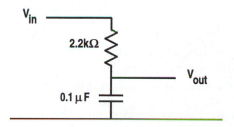

Figure 15.2: Example 15.1.

The network response is given by:

$$\frac{V_{out}}{V_{in}} = \frac{Z_2}{Z_1 + Z_2}$$

$$= \frac{\frac{1}{j2\pi fC}}{R + \frac{1}{j2\pi fC}}$$

$$= \frac{1}{1 + j2\pi fCR}$$

First calculate $\quad \tan\phi = -2\pi fCR$

$$= -2\pi 1500 \times 0.1 \times 10^{-6} \times 2200$$

$$= -2.07$$

So that $\quad \phi = \tan^{-1}(-2.07)$

$$= -1.12\,\text{rad or} - 64.3°$$

and the attenuation is $\quad \left|\dfrac{V_{out}}{V_{in}}\right| = \dfrac{1}{\sqrt{1 + \tan^2\phi}} = \dfrac{1}{\sqrt{1 + 2.07^2}}$

$$= 0.435 = 20\log 0.435\,\text{dB} = -7.23\,\text{dB}$$

If this circuit is constructed and the input and output waveforms are displayed on an oscilloscope then a trace similar to that in Figure 15.3 should be obtained.

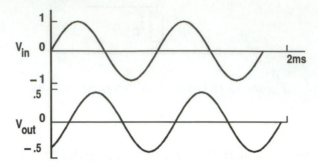

Figure 15.3: Input and output voltage waveforms for Example 15.1.

In the calculations we obtained a phase shift of -1.12 radians. In the oscilloscope diagram it can be seen that the output waveform is displaced to the right by 1.12 radians or 64° relative to the input voltage waveform. So we obtain the useful rule that:

- If the phase shift is positive then the output waveform is shifted to the left and is said to lead the input waveform.

- If the phase shift is negative then the output waveform is shifted to the right and is said to lag the input waveform.

15.2   Problems

15.1 Write down the expression for the output voltage waveform in Example 15.1 shown in Figure 15.3.

15.2 Calculate the attenuation and phase shift for the RC circuit shown in Figure 15.4 when $f = 500\,\text{Hz}$, $C = 22\,\text{nF}$ and $R = 10\,\text{k}\Omega$. Sketch the input and output voltage waveforms showing the amplitude and phase of the signals. Assume that the input signal is 1V_{pp}.

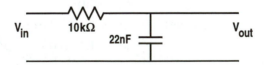

Figure 15.4: Problem 15.2.

15.3 Calculate the attenuation and phase shift for the CR circuit shown in Figure 15.5 when $f = 1.5\,\text{kHz}$, $C = 0.1\,\mu\text{F}$ and $R = 1.2\,\text{k}\Omega$. Sketch the input and output voltage waveforms showing the amplitude and phase of the signals. Assume that the input signal is of amplitude $1\,\text{V}$.

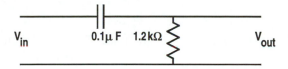

Figure 15.5: Problem 15.3.

15.4 Calculate the attenuation and phase shift for the LR circuit shown in Figure 15.6 when $f = 7\,\text{kHz}$, $L = 10\,\text{mH}$ and $R = 680\,\Omega$. Sketch the input and output voltage waveforms showing the amplitude and phase of the signals. Assume that the input signal is $1\,\text{V}_{\text{pp}}$.

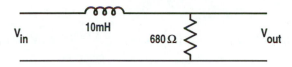

Figure 15.6: Problem 15.4.

15.5 Calculate the attenuation and phase shift for the RL circuit shown in Figure 15.7 when $f = 60\,\text{kHz}$, $L = 1\,\text{mH}$ and $R = 470\,\Omega$. Sketch the input and output voltage waveforms showing the amplitude and phase of the signals. Assume that the input signal is $1\,\text{V}_{\text{pp}}$.

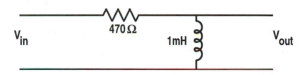

Figure 15.7: Problem 15.5.

15.6 Calculate the frequency for which the attenuation is $3\,\text{dB}$ for the circuit in Figure 15.8. Calculate the phase shift in degrees at this frequency.

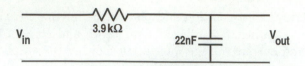

Figure 15.8: Problem 15.6.

15.7 Show that the bridge circuit in Figure 15.9 will be in balance, that is $V_A = V_B$, when $R_X = \frac{R_3 \times R_2}{R_1}$.

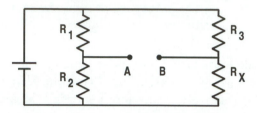

Figure 15.9: Problem 15.7.

15.8 Show that the Simple bridge in Figure 15.10 will be in balance when:

$$R_X = \frac{R_2 \times R_3}{R_1} \quad \text{and} \quad C_X = \frac{R_1}{R_3} \times C_2$$

Does the balance depend on the frequency of the voltage across the bridge? Note that the real and complex parts of the impedance equation must balance separately.

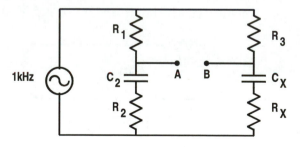

Figure 15.10: Problem 15.8.

Unit 16 Bode plot and frequency response

- The corner frequency, f_c, in hertz for a first order RC or RL filter is given by:
$$f_c = \frac{1}{2\pi CR} \quad \text{or} \quad \frac{R}{2\pi L}$$

- The filter response is approximated, on a dB versus log frequency plot, by two straight lines, one of slope 0 and the other of slope $\pm20\,$dB per decade and drawn through the point at the corner frequency and $0\,$dB, $(\log f_c,\ 0\,$dB$)$.

- The attenuation at the corner frequency is $-3\,$dB and the phase shift is $\pm\frac{\pi}{4}$ radians or $\pm45°$.

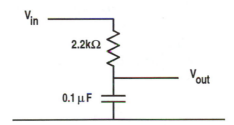

Figure 16.1: First order RC filter.

The frequency response of a filter such as that shown in Figure 16.1 is obtained by repeating the calculation of the attenuation carried out in Example 15.1 for a range of frequencies. When the response in decibels, dB, as a function of the log of the frequency, is plotted, a response curve similar to that shown in Figure 16.2 is obtained.

In doing these calculations the frequencies used were 10, 20, 50, 100, 200, 500, 1000, 2000, 5000, 10,000 Hz and the resulting points are shown on the plot. You should note that the sequence 1, 2, 5 and multiples of 10 give roughly equally spaced points on a log frequency plot. In the laboratory you should try to use these multiples where possible.

There is one special frequency for first order filters which is given by $f_c = \frac{1}{2\pi RC}$ Hz and is called the corner frequency. For this example the corner

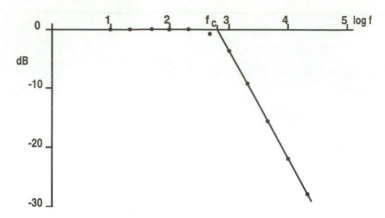

Figure 16.2: Amplitude response of first order filter.

frequency $f_c = \frac{1}{2\pi 2200 \times 0.1 \times 10^{-6}} = 724\,\text{Hz}$. Then $\log(724) = 2.86$ and this is marked on the log frequency axis as f_c.

If the point ($\log f_c$, 0 dB) is plotted, then the response curve can be approximated by two straight lines through this point as shown on the plot. One of the lines is at a constant 0 dB and is the frequency axis. The second line has a slope of -20 dB per decade, that is it drops by 20 dB for each decade in frequency or each change of 1.0 on the log frequency scale. The powerful feature of this Bode plot method is that you only need one number, the corner frequency, to be able to give a reasonably accurate plot of the frequency response of an RC or RL circuit. You do not need to do, for each frequency, the detailed computations which were done in order to obtain the points plotted in Figure 16.2.

At the corner frequency $\tan\phi = 2\pi f_c CR = 1$ and therefore the attenuation is:

$$\frac{1}{\sqrt{1 + \tan^2\phi}} = \frac{1}{\sqrt{1+1}} = \frac{1}{\sqrt{2}} = 0.707$$

Expressed in dB this becomes $20\log 0.707 = -3\,\text{dB}$ so a small rounding off of the sharp corner made by the straight lines at the intersection accommodates this $-3\,\text{dB}$ error at the corner frequency.

The second half of the Bode plot is the phase response. We have seen that the phase shift is $-45°$ at the corner frequency. The phase shift is $0°$ for frequencies where the response curve is flat. Where the response is falling at -20 dB per decade, the phase shift is approximately $\pm 90°$, depending on whether an RC or a CR filter is being used, so we have the approximate response as shown in the phase plot in Figure 16.3. This is a straight line

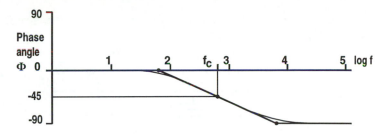

Figure 16.3: Phase plot.

approximation from 0° and either $0.1f_c$ or $10f_c$ and a second point at 45° and f_c.

When the amplitude response and the phase response are combined on one diagram with the log frequency scale the pair of curves is called a Bode plot. The full Bode plot is shown in Figure 16.4.

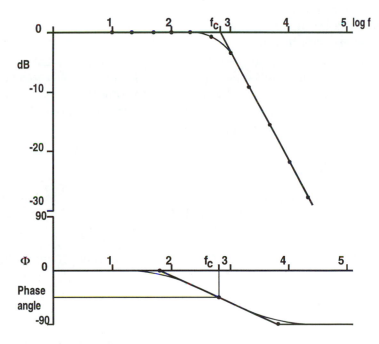

Figure 16.4: Bode plot for RC low pass filter.

The RC circuit which we have discussed is a low pass circuit. A CR circuit where the capacitor and resistor are interchanged is a high pass circuit. When you have an RC circuit or a CR circuit the type of response can be easily determined by remembering that at low frequencies a capacitor is essentially an open circuit and at high frequencies a capacitor is a short circuit. These approximations are illustrated in Figure 16.5.

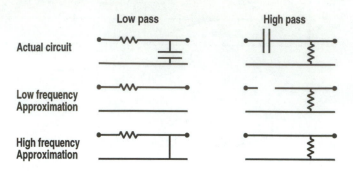

Figure 16.5: Extreme responses of filters.

We have discussed the Bode plots in terms of electronic circuits but Bode plots also have extensive applications in the analysis of instrumentation and control systems. For instance, a thermometer may have a response time constant of 30 seconds. This gives a corner frequency of $f_c = \frac{1}{2\pi 30} = 5.3\,\text{mHz}$. Such a thermometer would only be of use in situations where the temperature varies slowly as it would be necessary to wait for at least three time constants, $3 \times T = 90\,\text{seconds}$, to allow the temperature to stabilize before the thermometer gives a valid measurement of temperature.

16.1   Problems

16.1 Calculate the response of the filter shown in Figure 16.1 for the frequencies 10 Hz, 724 Hz, 1 kHz and 10 kHz and verify the response curve shown in Figure 16.2.

16.2 Calculate the attenuation at frequencies of 100, 1000, 10,000, 100,000 Hz for the circuits shown in Figure 16.6 and verify that the attenuation on the falling part of the response curves changes by 20 dB for each factor of 10 in frequency.

Figure 16.6: Problem 16.2.

16.3 Construct the Bode plot for the circuit shown in Figure 16.7.

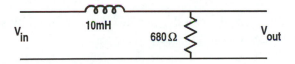

Figure 16.7: Problem 16.3.

16.4 Construct the Bode plot for the circuit shown in Figure 16.8.

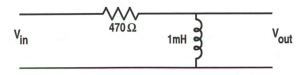

Figure 16.8: Problem 16.4.

16.5 Construct the Bode plot for the circuit shown in Figure 16.9.

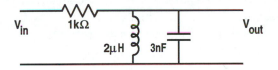

Figure 16.9: Problem 16.5.

16.6 When the heat flux incident on a radiant heat detector is modulated at 12 Hz, it is found that the output voltage is 0.707 of the output voltage when there is no modulation. Sketch the frequency response (Bode plot) for the heat detector.

Unit 17 Filter classification

In general, most filters can be specified by selecting one property or characteristic from each of the following groups:

- Active or passive

- High pass, low pass, band pass, band stop

- First, second, third ... order

- Smoothness of response (phase or amplitude).

Active and passive filters. So far we have only looked at passive filters which are filters constructed from passive components such as resistors, capacitors and inductors. In Unit 48 we will look at active filters which use transistors or operational amplifiers and usually have some gain.

High pass filters are characterized by the fact that they pass signals at high frequencies and attenuate or block signals at low frequencies.

Low pass filters are characterized by passing direct voltages and low frequencies and tend to attenuate high frequency signals. In general, they have a resistive connection between input and output.

Band pass filters pass signals at frequencies within a specified pass band and tend to block signals at frequencies above and below this pass band. They are usually specified as having a centre frequency and a 3 dB bandwidth or frequency range within which the response is within 3 dB of the response at the centre frequency. The Q factor of a filter is defined by:

$$Q = \frac{\text{Centre frequency}}{\text{Pass band}} = \frac{f}{B}$$

High Q factors are associated with sharp peaks in the frequency response which would be characteristic of a radio receiver selecting one signal frequency out of the many in the radio spectrum. There is a price to be paid for high Q factor or high selectivity and it is that the stability is reduced and the filter centre frequency tends to drift. As a general rule, passive band pass filters will usually contain both inductors and capacitors in the circuit.

Band stop filters block signals in a specified range or more accurately they attenuate the signals by some minimum amount usually specified in dB.

These band stop filters can be constructed from resonant LCs as shown in Figure 17.1. There is also another type of band stop filter which uses a low pass filter and a high pass filter operating in parallel. The band stop is then associated with the region between the two nonoverlapping filters where both responses attenuate the signal.

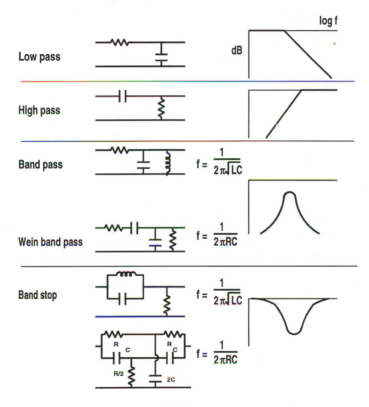

Figure 17.1: Typical filter circuits and response curves.

Filter order is most easily specified as the number of simple RC or RL filters which are combined in the composite filter. The order of the filter has its greatest effect on the slope of the shoulders of the filter response. If N is the order of the filter, then the total slope of the shoulder is $N \times 20\,\text{dB}$ per decade. The total slope means that two CR high pass filters in series will give a fall-off of the response curve of $40\,\text{dB}$ per decade of frequency. However, a CR high pass and RC low pass filter, when combined to give a band pass Wein filter, have a $20\,\text{dB}$ per decade fall-off on the low side and another $20\,\text{dB}$ per decade fall-off on the high side to give a total distributed fall-off of $40\,\text{dB}$ per decade.

Smoothness of response describes the permissible ripple or variation in the amplitude or phase shift of the signal within the filter pass band. It is usually specified by requiring the filter to be a Butterworth, Chebyshev, Bessel or some other specific design.

17.1 Example

17.1 Select a circuit and calculate suitable component values for a band pass filter having a centre frequency of 11 kHz.

From Figure 17.1, we can select a suitable LC filter circuit, as shown in Figure 17.2. The R_2 is included to denote the resistance of the inductor but we will ignore R_2 in our calculations since it is usually very small.

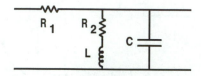

Figure 17.2: Circuit for band pass filter.

The readily available off the shelf component values for inductances are restricted in range so we will select the inductor first. Take a value of 1 mH as a reasonable first guess. Then:

$$\text{Using} \quad f_r \;=\; \frac{1}{2\pi\sqrt{LC}}$$

$$\text{we get} \quad C \;=\; \frac{1}{4\pi^2 f_r^2 L}$$

$$= \frac{1}{4 \times 9.87 \times 1.21 \times 10^8 \times 1 \times 10^{-3}}$$

$$= \frac{1}{4.777 \times 10^6}$$

$$= 0.21 \times 10^{-6}\,\text{F} = 0.21\,\mu\text{F}$$

17.2 Problems

17.1 Figure 17.3 shows a circuit for a band stop or notch filter. It is composed of a high pass filter in parallel with a low pass filter. Identify and sketch the low pass part of the filter. Identify and sketch the high pass part of the filter.

Plot the frequency response of each of these two component filters on the same sheet and construct the composite response. Plot the phase response of each of the component filters and construct the composite phase response.

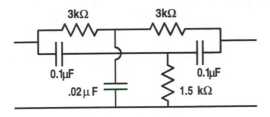

Figure 17.3: Problem 17.1.

17.2 An oscilloscope is switched into XY mode and the input signal to a filter is fed to the X deflection oscilloscope input and the filter output is fed to the Y deflection oscilloscope input as shown in Figure 17.4. After suitable adjustment of the oscilloscope sensitivity, an ellipse is obtained on the screen as shown in Figure 17.4.

A is the separation of the two points where the ellipse intercepts the Y axis and B is the vertical separation of the maximum and minimum Y displacements. Show that $\sin\phi = \frac{A}{B}$.

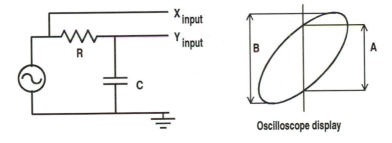

Figure 17.4: Problem 17.2.

17.3 Design a high pass filter having a corner frequency of 2.5 kHz and construct the Bode plot for the filter.

17.4 The RCL band stop filter shown in Figure 17.5 can be considered as an RC low pass filter at low frequencies and as an RL high pass filter at high frequencies. Show that the phase response of the filter is of the form shown in the diagram in Figure 17.5. Indicate the shape of the ellipse which you would expect to obtain for frequencies in each section of the phase response curve when the phase display circuit of Problem 17.2 is used.

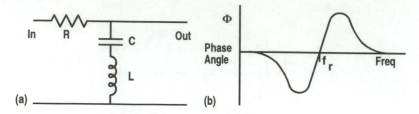

Figure 17.5: Problem 17.4.

17.5 Design a band pass filter for a centre frequency of 7.5 kHz and sketch the response.

17.6 Design a Wein filter for a centre pass frequency of 3.3 kHz.

17.7 If the component values for the circuit in Figure 17.2 are $R_1 = 1\,\text{k}\Omega$, $R_2 = 0\,\Omega$, $L = 20\,\text{mH}$ and $C = 0.5\,\mu\text{F}$, calculate the band pass centre frequency.

Unit 18 Fourier series

- Any repetitive waveform can be synthesized from the sum of sinusoidal waves of appropriate amplitude and phase.

- The frequencies of the Fourier components are the fundamental frequency and integer multiples of this frequency.

- The sharper the corners in the original waveform, the greater will be the amplitudes of the higher frequency Fourier components of the waveform.

- The response of any filter to a repetitive waveform is obtained by summing the responses for each of the Fourier components of the input waveform.

Analysis of waveforms

As this is an introductory course, we will discuss the Fourier analysis of waveforms in graphical terms rather than use the full mathematical treatment which is readily available in any text on Fourier series.

Take a sinusoidal wave of fundamental frequency, f_0, and amplitude 1 as shown in Figure 18.1 (a). Add to this waveform a sinusoid of frequency $3f_0$, the third harmonic, which has an amplitude of 33% of the fundamental. This is shown in Figure 18.1 (b) with the sum of the two waveforms shown in Figure 18.1 (c). Add to this sum a sinusoid of frequency $5f_0$ and amplitude 20% of the fundamental to get Figures 18.1 (d) and (e). Visualize this process continuing for all of the odd harmonics of f_0 given by $f = (2n + 1)f_0$ and having amplitudes $\frac{1}{2n+1}$, where n is an integer, eventually leading to the composite synthesized square waveform in Figure 18.1 (f).

As higher harmonics are added in, the corners of the square waveform are sharpened up as shown in Figures 18.1 (a), (c) and (e).

Thus we can see that a square waveform of period T can be considered as the sum of a sinusoidal waveform of fundamental frequency $f_0 = \frac{1}{T}$ combined with sinusoids at the odd harmonics of this fundamental frequency.

In our example we have taken the phase shifts of the harmonics to be zero, that is all of the harmonics are zero at times $0, \frac{T}{2}, T, \frac{3T}{2}, \ldots$ If the

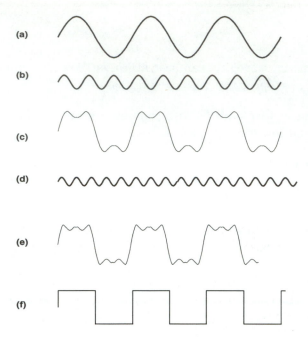

Figure 18.1: Construction of square waveform from Fourier components.

phase of the harmonics is not zero then the waveform synthesized can be quite different even though the amplitudes of the harmonics are unchanged.

Synthesis of filter response

In order to determine the effect of a filter on an arbitrary repetitive waveform, follow the following procedure:

- Obtain the Fourier spectrum of the input waveform.

- Calculate the effect of the filter on each of the Fourier components.

- Combine the modified components to obtain the output waveform.

This procedure can be carried out numerically but often the following graphical method will permit a rapid estimation of the output waveform to be obtained without a long calculation.

Plot the log of the amplitude of each of the Fourier components against the log of the frequency to get a diagram such as that shown in Figure 18.2 (a) which represents the frequency spectrum of the square wave. The small circles indicate the amplitudes of the Fourier component at that frequency.

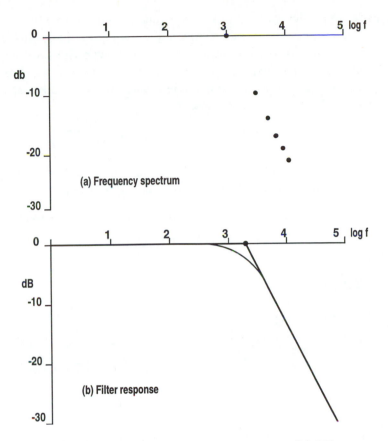

Figure 18.2: (a) Fourier spectrum of square wave (b) Filter response.

If the waveform having this frequency spectrum is passed through a filter with a corner frequency at 2 kHz which has a response curve or Bode plot such as that in Figure 18.2 (b) then the lower frequency components will emerge unchanged but the higher frequencies will be attenuated resulting in the spectrum shown in Figure 18.3.

At each frequency we have multiplied the amplitude of the Fourier component at that frequency by the magnitude of the attenuation of the filter to get the magnitude of the output. This is the powerful feature of the Bode plot approach. Since the log of the amplitude is plotted, all we have to do is, at each frequency, to **add** the logs of the signal and filter responses to get the log of the filter output and thence a log spectrum of the output.

This operation is most easily carried out if the amplitude of the input spectral components and the response curves for the filter(s) are all plotted on a single graph. The spectrum of the output is then obtained by a graphical adding of the signals at each frequency as is shown in Figure 18.3 in which the amplitudes of the Fourier components of the output are indicated by ×. For example, at a frequency of 3 kHz or at 3.48 on the log f axis, the filter response is -4 dB and the amplitude of the Fourier component of the square wave is -9 dB which gives the output:

$$-9\,\mathrm{dB} - 4\,\mathrm{dB} = -13\,\mathrm{dB}$$

as indicated by the arrows in Figure 18.3. Remember that the amplitudes are smaller than the reference and therefore the dB values are negative.

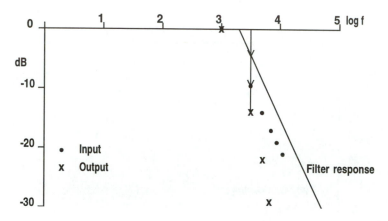

Figure 18.3: Each Fourier component is attenuated by the filter response at that frequency.

This example which we have just discussed represents what happens when a square wave is passed through an *RC* low pass filter which has a corner frequency which is close to the fundamental frequency of the square wave. The circuit and input and output waveforms are shown in Figure 18.4 and it can be seen that the higher frequency harmonics associated with the sharp edges of the square waveform have been attenuated by the filter to leave a much smoother output waveform.

In our discussion, we have not mentioned the phase shifts which occur in the filter and how they may affect the output waveform. If the filter response is such that the phase delay in the filter is constant over the pass band of the filter then the distortion of the waveform due to phase changes will be minimized.

Also the sharper the corners in the original waveform, the greater will be the amplitudes of the higher frequency components of the waveform.

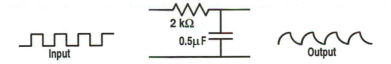

Figure 18.4: Distortion of a square wave by a low pass filter.

18.1 Problems

18.1 Estimate by graphical summation the relative amplitudes of the first two Fourier components of the triangular waveform shown in Figure 18.5.

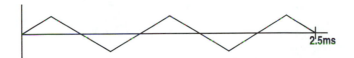

Figure 18.5: Triangular waveform for Problem 18.1.

18.2 Estimate by graphical summation the relative amplitudes and phases of the first four components of the sawtooth waveform shown in Figure 18.6.

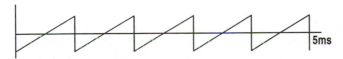

Figure 18.6: Sawtooth waveform for Problem 18.2.

18.3 Sketch a circuit for a CR high pass filter with a corner frequency of 2 kHz. A square wave of fundamental frequency 1 kHz is passed through this filter. Sketch the spectrum of the square wave, the response of the filter and the spectrum of the output waveform. Sketch the shape of the distorted output waveform.

18.4 A square waveform of fundamental frequency 20 kHz is passed through a band pass filter which has a centre frequency of 100 kHz and a 3 dB bandwidth of 15 kHz. Which Fourier components of the square waveform will be passed through the band pass filter? Sketch the output voltage waveform.

Unit 19 Thévenin's theorem

- Any linear electronic system, having two output terminals, can be fully modelled by a voltage source, V_S, in series with an impedance, called the output impedance, Z_{out}.

- A voltage source gives a constant output voltage which is independent of the current drawn from the voltage source.

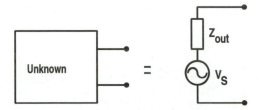

Figure 19.1: Thévenin equivalent circuit.

The voltage sources used in Thévenin analysis of circuits are devices which give a constant voltage at their output terminals. In the case of DC voltage sources, the voltage is constant and independent of the current which is drawn from the voltage source. In the case of AC voltage sources, the amplitude of the output voltage remains constant and the voltage varies sinusoidally at some fixed frequency. The Thévenin analysis method is the same for DC and AC circuits except that the complex impedance is used when capacitors or inductors are included in the circuit. We will therefore not make any fundamental distinction between the analysis of DC and AC circuits.

The concept of a voltage source is an idealization but there are a number of real examples which approximate to the ideal. Two good examples which approximate to the ideal voltage source are:

- The mains AC power supply. The RMS voltage at the wall socket remains at a constant 240 V independently of whether we draw no current or draw a $\frac{1}{4}$ A when we connect a 60 W bulb or draw 8 A when we connect a 2 kW electric fire.

- A lead acid car battery. The voltage at the battery terminals in a car remains constant at 12 V whether we operate the clock which draws 10 mA, the radio which draws 3 A, the headlights which draw 14 A or the starter motor which draws 120 A.

As an example of the Thévenin analysis method we consider how we can obtain the Thévenin model for a battery comprised of a number of dry cells in series.

If the output voltage from a fresh PP9 type battery is measured, a value of about 9 V is obtained. The voltmeter has a high resistance and therefore does not draw any significant current from the battery. There is therefore no voltage drop across the internal resistance of the battery and the voltmeter reading is then equal to that of the Thévenin voltage source.

If an ammeter is connected across the battery terminals, the battery is effectively shorted but an infinite current is not obtained. Typically the current will be 3 A. (At least for a short time until the battery becomes discharged!) The current is limited by the internal resistance of the electrolyte and electrodes of the battery.

We can model the PP9 battery by a 9 V voltage source and a resistance internal to the battery which is called the output resistance, R_{out}, which is physically equivalent to the resistance of the internal electrolytes and electrodes. We can then model these two measurements of the open circuit output voltage and the short circuit current with the circuits in Figure 19.2.

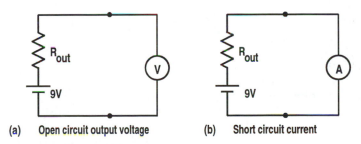

(a)   **Open circuit output voltage**   (b)   **Short circuit current**

Figure 19.2: Measurement of Thévenin equivalent circuit of a PP9 battery.

From this simple model, we can see that:

$$R_{out} = \frac{V_{out\ open\ circuit}}{I_{out\ short\ circuit}}$$

For the PP9 battery example we therefore have $R_{out} = \frac{9\,\text{V}}{3\,\text{A}} = 3\,\Omega$.

In general, we do not normally short the output of a circuit with an ammeter in order to measure the short circuit current. The concept of short circuit current is, however, very useful for defining the output resistance.

If we return to the examples of voltage sources at the start of the discussion, we can say that the 240 V AC mains and the 12 V car battery voltage sources are really Thévenin sources in which the R_{out} is very small. If, for instance, a car battery had an output resistance of 1 Ω then the maximum current which it could drive through an external circuit would be $\frac{12\,\text{V}}{1\,\Omega} = 12\,\text{A}$ which would not even light the car headlights let alone operate the starter motor!

The output resistance of an electronic system can be measured by measuring the voltage at the output terminals as a function of the current drawn from the terminals. The short circuit current can then be obtained by extrapolation. It is in general not good practice to short out the output of an electronic system in order to measure the short circuit current, not only because of the possibility of damaging the electronics but also because many electronic circuits have special circuits included which detect a short circuit and close down the output before any damage is done. The electronic system is no longer linear and therefore one of the basic conditions for applicability of Thévenin's theorem no longer applies.

A circuit suitable for measuring the output resistance of an electronic system is shown in Figure 19.3 (a).

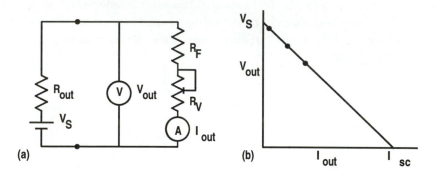

Figure 19.3: Practical measurement of Thévenin equivalent circuit.

The fixed resistor, R_F, controls the maximum current which can flow, since $I_{max} = \frac{V_S}{R_{out}+R_F}$. This I_{max} is specified by the manufacturer of the electronic system so as not cause any damage to the system. The variable resistor, R_V, allows the current to be varied so as to obtain a number of points on the V–I curve as shown in Figure 19.3 (b). The line through these points is then extrapolated to obtain the short circuit output current, $I_{short\ circuit}$.

It is not necessary to extrapolate to the short circuit current in order to obtain the output resistance as it is easily seen that the slope of the V–I characteristic gives the output resistance.

19.1 Examples

19.1 The table below shows the terminal voltages and currents which were measured when the specified resistance was connected across a PP3 battery. The circuit used is shown in Figure 19.4 (a). Calculate the Thévenin equivalent circuit for the PP3 battery.

Resistance	Output voltage V	Output current mA
$500\,\text{k}\Omega$	8.91	0.00
$1\,\text{k}\Omega$	8.82	9.18
$470\,\Omega$	8.75	18.70
$180\,\Omega$	8.56	49.10
$100\,\Omega$	8.39	85.70

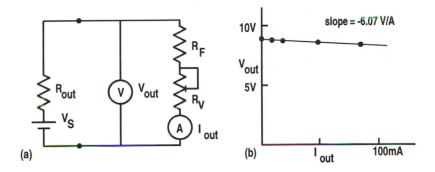

Figure 19.4: Example 19.1.

The output voltage is plotted against current as shown in Figure 19.4 (b). The slope of the curve is $-6.07\,\text{VA}^{-1}$ which, when extrapolated, gives an intercept at $I_{short\ circuit} = \frac{8.91}{6.07} = 1.47\,\text{A}$ giving:

$$R_{out} = \frac{V_{open\ circuit}}{I_{short\ circuit}} = \frac{8.91}{1.47} = 6.07\,\Omega$$

which is the slope of the V–I characteristic! So this PP3 can be modelled by an 8.91 V voltage source in series with $6.07\,\Omega$.

19.2 Calculate the Thévenin equivalent of the circuit shown in Figure 19.5.

A voltmeter draws very little current from a circuit because of its high input resistance of $10\,\text{M}\Omega$, so connecting a voltmeter across the output does not cause any significant current to flow in the $470\,\Omega$ resistor and there is then no significant voltage drop across the $470\,\Omega$. Therefore the

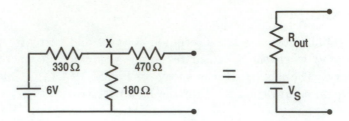

Figure 19.5: Example 19.2.

voltage at point X is the same as the voltage at the output terminals. We now have a potential divider of 330 Ω and 180 Ω in series which gives an open circuit output voltage of:

$$V_S = V_{open\ circuit} = \frac{180}{330 + 180} \times 6\,\text{V} = 2.12\,\text{V}$$

If an ammeter is connected across the output, the circuit shown in Figure 19.6 (b) results. Remember that an ammeter has an input resistance approaching zero ohms.

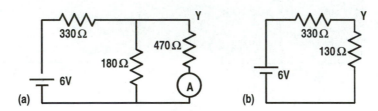

Figure 19.6: Example 19.2. Short circuit current.

The 180 Ω is now in parallel with the 470 Ω to make 130 Ω and this gives the voltage at point Y as:

$$V_Y = \frac{130}{330 + 130} \times 6\,\text{V} = 1.7\,\text{V}$$

This is the voltage across the 470 Ω and therefore the short circuit output current is $\frac{1.7}{470} = 3.61\,\text{mA}$.

The output resistance for the Thévenin equivalent circuit is then:

$$R_{out} = \frac{V_{open\ circuit}}{I_{short\ circuit}} = \frac{2.12\,\text{V}}{3.61\,\text{mA}} = 586\,\Omega$$

We can then replace the circuit of Figure 19.6 (a) by its Thévenin equivalent as shown in Figure 19.7.

Figure 19.7: Example 19.2. Thévenin equivalent circuit.

19.3 Calculate the voltage which would be measured by a 10 MΩ input impedance voltmeter connected, as shown, between ground and point A in the circuit in Figure 19.8.

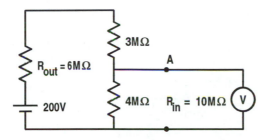

Figure 19.8: Example 19.3.

The input impedance of the voltmeter gives a potential divider lower arm of 4 MΩ in parallel with 10 MΩ, that is 2.86 MΩ. The upper arm of the potential divider is the output resistance of the voltage source, 6 MΩ in series with 3 MΩ.

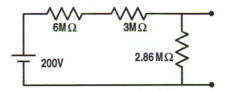

Figure 19.9: Effect of meter loading.

This effective circuit is shown in Figure 19.9. The voltage which is indicated by the voltmeter is then:

$$\frac{2.86}{2.86 + 6 + 3} \times 200\,\text{V} = 48.2\,\text{V}$$

You should note that connecting a voltmeter to the circuit can change the voltages at various points in the circuit. This is called **measurement loading**. It is usually only a significant problem when the resistances in the circuit are large (of the order of megohms).

19.2   Problems

19.1 The following measurements were made of the output terminal voltages and currents for an electronic circuit. Calculate the Thévenin equivalent for the circuit.

Voltage V	Current mA
16.0	0
15.8	320
15.3	1100
14.9	1770

19.2 Calculate the Thévenin equivalent of the circuit shown in Figure 19.10.

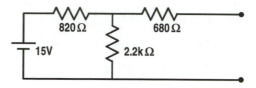

Figure 19.10: Problem 19.2.

19.3 The 6 V battery in Example 19.2 is replaced by a 9 V battery. Calculate and sketch the new Thévenin equivalent circuit.

19.4 Calculate the Thévenin equivalent for the circuit shown in Figure 19.11.

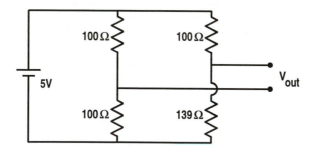

Figure 19.11: Problem 19.4.

19.5 Calculate the Thévenin equivalent for the circuit shown in Figure 19.12.

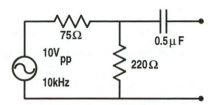

Figure 19.12: Problem 19.5.

19.6 Calculate the true voltages at points A and B in the circuit diagram shown in Figure 19.13. Calculate the voltages, relative to ground, which would be measured with a voltmeter having a $10\,\text{M}\Omega$ input resistance. You may assume that the $1000\,\text{V}$ power supply has a negligible output resistance.

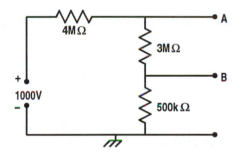

Figure 19.13: Problem 19.6.

Unit 20 Norton's theorem

- Any linear two terminal electronic system can be fully modelled by a current source, I_S, in parallel with a shunt impedance Z_{out}.

- A current source drives a constant current through any circuit connected to it. The circuit symbol used for a current source is two intersecting circles as shown in the diagram.

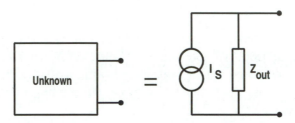

Figure 20.1: Norton equivalent circuit.

This theorem is the complement of Thévenin's theorem but uses current sources and can therefore give an alternative analysis viewpoint for circuits. The one disadvantage is that examples of constant current sources are not as readily available as examples of voltage sources.

Perhaps the simplest way of viewing a current source is to visualize it as a high voltage source in series with a large series resistor, as shown in Figure 20.2.

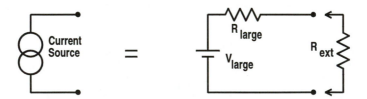

Figure 20.2: Current source.

For a wide range of externally connected resistors, the current will be essentially independent of the external resistance because if:

$$R_{large} \gg R_{ext} \quad \text{then} \quad \frac{V_{large}}{R_{large} + R_{ext}} \approx \frac{V_{large}}{R_{large}} = I_S$$

For comparison purposes, the characteristic curves for a voltage source and for a current source are shown in Figure 20.3.

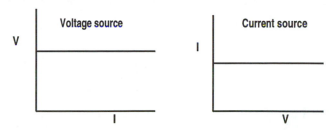

Figure 20.3: *I–V* characteristics of voltage and current sources.

In modelling circuit devices and systems, either the Thévenin or the Norton approach can be taken but the characteristics of the device being modelled should be examined and if the characteristics approximate to those of a voltage source (see Figure 20.3) then the Thévenin method is appropriate. If the device characteristic approximates the current source, then the Norton approach is more appropriate.

Examples of devices and systems where the Thévenin approach is appropriate are: batteries, diodes, DC constant voltage power supplies, audio and RF signal generators, lead acid battery chargers. Examples of suitable applications of the Norton analysis approach are: photodiodes in reverse bias, output characteristics of transistors and FETs in amplifiers, NiCd battery chargers, constant current supplies for fluorescent lamps and laboratory spectral lamps.

We will see in the worked examples that it is easy to convert from one model to another so it is possible to model circuit systems quite effectively using either method. However, some insight into the circuit system is lost if an inappropriate model is used.

There is frequently a control panel adjustment for the Thévenin voltage or the Norton current source which allows the operator to set the output to the required value. Once set, the system then gives constant output. You should take care with the usage of the terms variable, fixed and constant as applied to voltage and current sources.

In order to measure the Norton equivalent of any circuit, it is only necessary to measure the open circuit voltage and the short circuit current.

- The current source is given by: $I_S = $ Short circuit current.

- The output impedance is given by: $Z_{out} = \frac{V_{open\ circuit}}{I_{short\ circuit}}$

20.1   Examples

20.1 Convert from Thévenin to Norton equivalent circuit: calculate the Norton equivalent circuit of the Thévenin circuit shown in Figure 20.4.

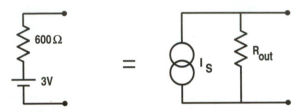

Figure 20.4: Example 20.1.

Short circuit current:

The Norton current source is given by:

$$
\begin{aligned}
I_S &= I_{short\ circuit} \\
&= \frac{V_{out}}{R_{out}} \\
&= \frac{3\,\text{V}}{600\,\Omega} \\
&= 5\,\text{mA}
\end{aligned}
$$

Open circuit voltage:

The output voltage from the Norton circuit is:

$$
\begin{aligned}
V_{out} &= V_{open\ circuit} \\
&= I_S \times R_{out} \\
\text{Therefore} \quad 3\,\text{V} &= 5\,\text{mA} \times R_{out} \\
R_{out} &= \frac{3\,\text{V}}{5\,\text{mA}} \\
&= 600\,\Omega
\end{aligned}
$$

20.2 Convert from Norton to Thévenin equivalent circuit: calculate the Thévenin equivalent of the Norton circuit shown in Figure 20.5.

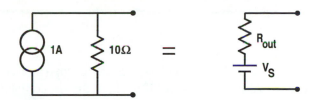

Figure 20.5: Thévenin equivalent of Norton circuit.

Open circuit voltage $V_{out} = I_S \times R_S = 1\,\text{A} \times 10\,\Omega = 10\,\text{V}$

Short circuit current $I_{short\ circuit} = \dfrac{V_S}{R_{out}} = 1\,\text{A}$

Therefore $\quad R_{out} = \dfrac{10\,\text{V}}{1\,\text{A}} = 10\,\Omega$

20.2 Problems

20.1 Calculate the Norton equivalent of the circuit shown in Figure 20.6.

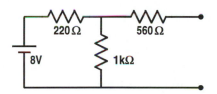

Figure 20.6: Problem 20.1.

20.2 Calculate the Norton equivalent of the circuit shown in Figure 20.7.

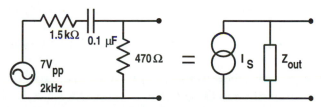

Figure 20.7: Problem 20.2.

Unit 21   Principle of superposition

- In a linear system, with several causes acting to give a combined effect, the net effect is determined by adding the individual effects.

- In electronic circuits, the individual effect of a voltage or current source is determined by removing all of the other voltage and current sources and replacing the voltage sources by short circuits and the current sources by open circuits.

The use of the principle of superposition can lead to major simplifications in the analysis of electronic circuits because most analog circuits are linear or at least they are linear in a small region near the operating point. The advantage of using analysis methods based on the principle of superposition is that a direct and proportional link is maintained between the application of an input and the effect at the output. This gives a significantly greater understanding of the circuit operation than the methods based on Kirchhoff type analysis where the link between cause and effect is frequently obscured by the computations.

You should, however, remember that there are limits to linearity. If one man can dig a certain size hole in 10 hours, two men can dig the same hole in five hours but 20 men could not dig it in half an hour because they would get in each others' way and the job is therefore no longer linear!

Suppose we need to obtain the current in R_3 in Figure 21.1 (a). The circuit is redrawn twice with only one voltage source remaining in each version (or 10 times if there are 10 voltage sources). The gaps where the voltage sources were located are shorted out and the current in R_3 due to each of the voltage sources is calculated as indicated in Figure 21.1 (b) and Figure 21.1 (c). The current which flows when both of the voltage sources are present is the algebraic sum of the currents which flow in R_3 when each of the voltage sources is present.

In dealing with problems using the principle of superposition it is important that you draw the modified circuit at each stage. It is also important that you indicate the defined direction of current flow on the diagram. There is no problem if the initial assignment of the current direction is incorrect as all that will happen is that the sign of the current flow will be reversed.

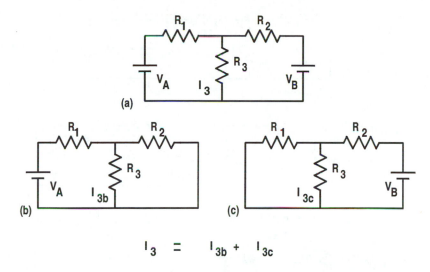

$$I_3 = I_{3b} + I_{3c}$$

Figure 21.1: Splitting of circuit before applying principle of superposition.

21.1 Examples

21.1 Calculate the current in the $8\,\Omega$ resistor in Figure 21.2 (a).

There are two voltage sources present so there are two sets of calculations to be carried out. These are indicated in the two vertical sets of circuits in Figures 21.2 (b) to (d) and (e) to (g).

In (b) the $10\,\mathrm{V}$ battery has been removed and replaced by a short.

In (c) the $3\,\Omega$ resistor has been removed since it does not influence the current in the $8\,\Omega$ resistor.

In (d) the $5\,\Omega$ and $8\,\Omega$ have been replaced by their parallel equivalent resistor $3.07\,\Omega$.

By application of the potential divider principle we calculate the voltage across the $3.07\,\Omega$. This is the same as the voltage across the $8\,\Omega$.

The voltage across the $8\,\Omega$ due to the $6\,\mathrm{V}$ source is:

$$V_{6,8} = \frac{3.07}{3.07 + 4} \times 6\,\mathrm{V} = 2.60\,\mathrm{V}$$

This gives the current in the $8\,\Omega$ resistor due to the $6\,\mathrm{V}$ source as:

$$I_{6,8} = \frac{2.60}{8} = 0.33\,\mathrm{A}$$

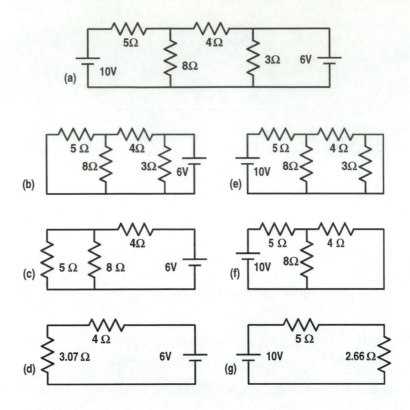

Figure 21.2: Calculate the effect of each cause separately.

On the right hand side of the diagram, replacement of the 6 V source by a short circuit gives Figure 21.2 (e).

The 3 Ω is shorted out and therefore has no effect. We then get the circuit in (f).

Combine 8 Ω and 4 Ω in parallel to get 2.66 Ω.

Using the potential divider, the voltage across the 2.66 Ω and therefore across the 8 Ω in (g) is:

$$V_{10,8} = \frac{2.66}{2.66 + 5} \times 10 = 3.48\,\text{V}$$

The current in the 8 Ω due to the 10 V source is then:

$$I_{10,8} = \frac{3.48}{8} = 0.43\,\text{A}$$

The total current in the 8 Ω is therefore $0.33 + 0.43 = 0.76$ A.

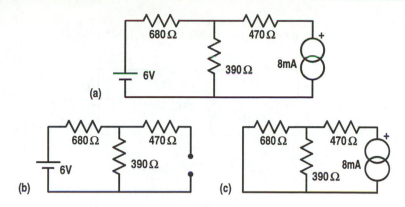

Figure 21.3: Example 21.2.

21.2 Calculate the current in the 390 Ω resistor in Figure 21.3 (a).

Redraw the circuit to eliminate the current source as shown in (b). The current source is replaced by an open circuit. Redraw the circuit to eliminate the voltage source as shown in (c). The voltage source is replaced by a short circuit.

In (b) the current in the 390 Ω due to the 6 V voltage source is:

$$I_{6,390} = \frac{6\,V}{680\,\Omega + 390\,\Omega} = 5.6\,\text{mA}$$

In (c) the voltage source has been eliminated and replaced by a short and we see that the 680 Ω is in parallel with the 390 Ω to give 248 Ω. The 8 mA passes through the 248 Ω and gives a voltage of $248 \times 0.008 = 1.98$ V which appears across the 390 Ω and 680 Ω in parallel.
The current in the 390 Ω due to the 8 mA current source is:

$$I_{8,390} = \frac{1.98}{390} = 5.1\,\text{mA} \quad \text{giving} \quad I_{total} = 5.6 + 5.1 = 10.7\,\text{mA}$$

21.2 Problems

21.1 Calculate the current in the 270 Ω resistor in Figure 21.4.

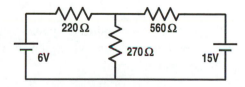

Figure 21.4: Problem 21.1.

21.2 Calculate the current in the $100\,\Omega$ resistor in Figure 21.5.

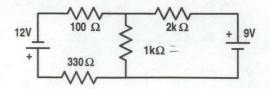

Figure 21.5: Problem 21.2.

21.3 Calculate the voltage across the $1\,\text{k}\Omega$ resistor in Figure 21.6.

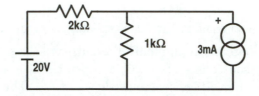

Figure 21.6: Problem 21.3.

21.4 Calculate the voltage wrt ground at point X in Figure 21.7.

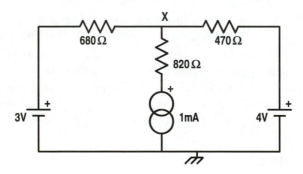

Figure 21.7: Problem 21.4.

21.5 Explain why voltage sources are replaced by short circuits and current sources are replaced by open circuits when the principle of superposition is used in the analysis of circuits.

Unit 22   Semiconductor materials

- Substances can be divided into three groups according to their electrical properties: conductors, semiconductors and insulators.

- The principal semiconductor materials used in electronics are the indirect band gap materials, silicon (Si) and germanium (Ge).

- The principal semiconductor materials used in optoelectronics and in very fast electronics are the direct band gap III–V semiconductors such as gallium arsenide (GaAs) and indium phosphide (InP).

- Intrinsic semiconductors are pure semiconductors in which the number of holes is equal to the number of mobile electrons.

- Extrinsic semiconductors have been doped to make either p-type material (in which the positive holes are the majority carriers and the electrons are minority carriers) or n-type material (in which the negative electrons are the majority carriers and the holes are minority carriers).

- Semiconductor equation: $n \times p = n_i^2 = $ constant for constant T. If the concentration of holes, p, increases then the concentration of electrons, n, decreases in proportion.

- Metallic conductors: resistance increases with increasing temperature.

- Semiconductors: resistance decreases with increasing temperature.

Consider the vibration of a stretched string such as a violin string. The string is constrained so that it is fixed at each end. The possible waves which can be present on the string correspond to an integral number of half wavelengths as shown in Figure 22.1.

In quantum mechanics, a particle such as an electron is described by a wave function. When an electron is constrained to a one dimensional box with rigid ends, the Schrödinger equation which describes the system has a similar mathematical form to the differential equation which describes the vibration of a string and which leads to discrete vibrational modes.

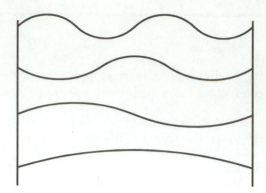

Figure 22.1: Vibrational modes on a stretched string.

However, the one significant difference is that the vibrational modes of the string can have any energy whereas a particular mode for a particle in a box has an energy which is given by $E_j = E_1 j^2$ where j is an integer, E_1 is the energy of the lowest energy mode and E_j is the energy of the jth mode. We can therefore plot the electron energy as a function of the mode number as shown in Figure 22.2 where the discrete modes lie on the parabola, $E_j = E_1 j^2$, and are indicated by circles.

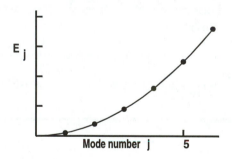

Figure 22.2: Plot of energy versus mode number.

Returning to the waves on a string, if we loosely constrain the string at the centre, then the large amplitude vibrations of the odd numbered modes, for which $j = 1$, $j = 3$, $j = 5$ etc., are inhibited. This is illustrated in Figure 22.3. The constraint acts to 'forbid' a mode. Placing the constraint in different positions causes a different set of modes to be 'forbidden'.

Now consider the electron in a one dimensional box. When atoms are placed at regular intervals in the box to form a one dimensional crystal, some of the solutions or modes obtained for Schrödinger's equation will be

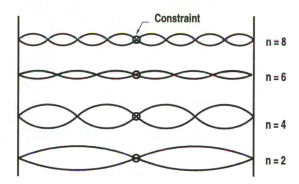

Figure 22.3: Effect of constraint in 'forbidding' modes.

forbidden. Essentially the box is sectioned by the atoms and within the crystal only electrons with certain wave vector values, k, can propagate through the crystal without being scattered by the atoms.

In a crystal, the quantized wave vector specifies the electron momentum, k, and the energy is given by $E(k) = \frac{h^2 k^2}{4\pi^2 2m^*}$ where h is Planck's constant and m^* is the effective electron mass. Also since waves propagate in both directions, both positive and negative values of k must be considered. The number of distinct modes or values of k is N where $N = \frac{l}{a}$, a is the lattice size (lattice constant) and l is the length of the crystal. The set of distinct modes is called a Brillouin zone.

When k exceeds the value of $\frac{\pi}{a}$, set by the size of the crystal lattice, the momentum values differ by multiples of $\frac{2\pi}{a}$ but are indistinguishable within the crystal due to interchanges of momentum between electrons and phonons (vibrational modes of the crystal lattice). However, the energy continues to increase along the parabolic curve and moves into a new Brillouin zone. Thus it is possible to have electrons which have the same momentum within the crystal but have significantly different energies. At the transition from one Brillouin zone to the next there is a significant deviation from the parabolic shape of the curve (a localized bending of the E–k curve) which leaves a band gap where there are no energy states and which results in a forbidden band. The resulting band structure is shown in Figure 22.4.

We have only considered a one dimensional crystal. In a real three dimensional crystal the electron wave functions can propagate along each of the principal axes of the crystal in the [100], [010] and [001] directions. There is also propagation in diagonal directions through the crystal shown as [110] and [210] directions in the two dimensional projection in Figure 22.5 or, for example, in the fully diagonal [111] direction in the 3D crystal. The spacing between the planes in the crystal is different for these diagonal directions,

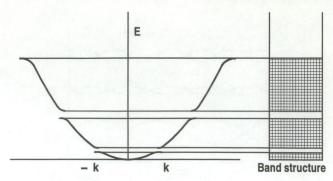

Figure 22.4: Energy levels and band structure.

leading to different and distinct E–k curves.

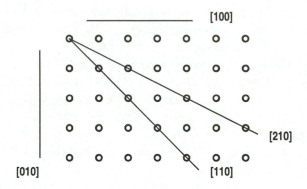

Figure 22.5: Directions in the crystal.

In a real crystal, the electrons in the lower atomic orbitals remain tightly bound to the individual atoms. However, the outer shell valence electrons, which interact with adjacent atoms, can no longer be considered to be attached to individual atoms but rather must be considered to be delocalized throughout the crystal with a distribution corresponding to the amplitudes of the wave functions which we have just discussed.

Since electrons are fermions, the Pauli exclusion principle applies and no two electrons can have the same set of quantum numbers so the available states fill up from the lowest energy states to a level called the Fermi level, E_F. At $T = 0$ K the distribution of electrons at the Fermi level cuts off sharply but at any higher temperature there is some rounding off of the electron distribution around the Fermi level due to thermal excitation and the probability of an electron having an energy E is then given by:

$$P(E) = \frac{1}{1 + e^{\frac{(E-E_F)}{kT}}}$$

We can now see how this band theory can be applied to explain the electrical properties of solids.

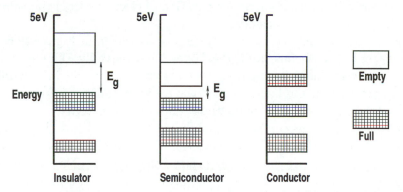

Figure 22.6: Band structure in insulators, semiconductors and conductors.

In insulators, as shown in Figure 22.6, the available electrons have filled the lower valence bands completely and there are no electrons in the next band up in energy. The energy gap or band gap between the highest full band and the lowest empty band is large, usually greater than 3 eV, and there is no mechanism for exciting significant numbers of electrons into the upper band where they could move between empty states and give electrical conduction through the crystal. The crystal is therefore an insulator because of the large energy gap.

In semiconductors, this energy gap between the top of the full valence band and the bottom of the empty conduction band is smaller, of the order of 1 eV. The thermal vibrational modes or phonons of the crystal have energies of kT. At room temperature, 293 K, we calculate:

$$kT = 1.38 \times 10^{-23} \times 293 = 4.04 \times 10^{-21} \, \text{joules} = \frac{4.04 \times 10^{-21}}{1.6 \times 10^{-19}} = 0.025 \, \text{eV}$$

The crystal vibrational modes can transfer energy to the electrons and excite the electrons across the energy band gap. However, only phonons at the extreme tail of the thermal phonon distribution are energetic enough to excite an electron across the band gap. Therefore the fraction of electrons excited to the conduction band is small. This small fraction multiplied by the large number of valence electrons in the crystal does give enough electrons in the conduction band to give significant electrical conduction. A small increase in the temperature gives a large increase in the fraction of electrons excited into the conduction band and therefore causes a large decrease in the resistivity of a semiconductor with increasing temperature—a negative temperature coefficient of resistance.

It should also be noted that the only difference between the insulators and the semiconductors is the size of the band gap energy and the consequent much lower probability of thermal excitation of electrons from the valence to the conduction band for insulators.

In the case of conductors, usually metals, as shown in Figure 22.6, there are only enough electrons to fill the upper band partially. This means that very little extra energy is required to move an electron up into a different energy state within the band so that the electron can move freely through the crystal and give a high conductivity and a low resistivity.

In conductors, the number of mobile electrons does not change significantly with temperature. When an electric field is applied across the crystal, the electrons accelerate and move through the crystal. The electrons also collide with the vibrating atoms of the crystal via electron-phonon interactions. These collisions randomize the imposed velocity due to the external electric field. The higher the temperature, the more phonons are present, the more frequent are the collisions and the more rapidly the electron velocities are randomized. This reduces the conductivity and increases the resistivity. Metallic conductors therefore have a positive temperature coefficient of resistance (TC of R). The sign of the temperature coefficient of resistance is one of the more important differences between semiconductors and metallic conductors.

The table shows the relative values for the resistivities of a number of representative materials. There is a factor of $\approx 10^9$ change in the resistivity in going from one group to the next.

Material	Resistivity, ρ, $\Omega\,\mathrm{m}$	Sign of TC of R
Glass	10^{10}	
Silicon	2000	−
Germanium	0.5	−
Copper	1.7×10^{-8}	+

In a pure or intrinsic semiconductor material, if an electron is excited up into the conduction band, there will be electrical conduction due to this movable electron. However, in semiconductors there is also another mechanism for conduction. The hole left in the valence band by the excited electron can be considered as a positive charge carrier since an adjacent valence electron can hop into the hole to give an effective movement of the hole in a direction opposite to the direction of movement of the electron which hopped. Therefore, we have p-type carriers or holes which can also move through the lattice.

Electrons and holes behave similarly and their behaviour is modelled using Fermi-Dirac statistics. The consequence is that the probability that a particular energy level, E, is occupied is given by the Fermi-Dirac distribution function:

$$P(E) = \frac{1}{1 + e^{\frac{(E-E_F)}{kT}}} \approx 1 - \exp\left(\frac{E - E_F}{kT}\right)$$

where E_F is the Fermi level energy. This exponential function leads to similar exponential functions appearing in the equation which describes variation of current as a function of applied voltage in a diode.

Extrinsic semiconductors result when silicon or germanium, which are Group IV elements in the Periodic Table, are doped with small quantities of either Group III elements such as boron atoms which act as acceptors for electrons to reduce the free electron concentration and give p-type semiconductor material, or Group V elements such as phosphorous atoms which act as donors of electrons to increase the free electron concentration and give n-type semiconductor material.

If n is the electron concentration, p is the hole concentration and n_i is the carrier concentration in the intrinsic material then it can be shown that:

$$n \times p = n_i^2 \qquad \text{(The semiconductor equation)}$$

at a constant temperature and after thermal equilibrium has been reached. Therefore when n-type semiconductor is formed by doping, the semiconductor equation shows that the concentration of p-type carriers in the n-type material is reduced below that which is found in intrinsic material. In n-type material, n-type carriers (electrons) are the majority carriers and p-type carriers (holes) are the minority carriers and conversely for p-type material.

It is possible to obtain concentrations other than those predicted by $n \times p = n_i^2$ over a local region by injecting, say, large numbers of p-type holes into an n-type region by applying appropriate voltages to some fabricated structure in the semiconductor crystal such as the base region in a transistor. If the injection process is stopped then the system returns to thermal equilibrium and $n \times p = n_i^2$ applies once more.

22.1   Problems

22.1 A thermistor is a semiconductor resistive device used for temperature sensing and measurement. The resistance R_2 at temperature t_2 K is given by the formula:

$$R_2 = R_1 e^{\left(\frac{B}{t_2} - \frac{B}{t_1}\right)}$$

where R_1 is the resistance at a reference temperature t_1 K and B is a constant for the particular type of thermistor.

A GM103 type thermistor has resistance $10\,\text{k}\Omega$ at $25\,°\text{C}$ and $B = 3555\,\text{K}$.

Calculate the resistance of a GM103 at each of these temperatures: $-6\,°\text{C}$, $30\,°\text{C}$, $50\,°\text{C}$ and $120\,°\text{C}$.

22.2 A constant **voltage** is applied along a resistive track formed on a silicon wafer. The resistance of the track at $20\,°\text{C}$ is R_s.

What is the formula for the power which is dissipated in the resistive track?
Will the temperature of the silicon increase or decrease?
Will the resistance of the resistive track increase or decrease?
Will the power dissipation increase or decrease?
Is the system thermally stable or is it subject to thermal runaway?

22.3 A constant **current** is driven through a resistive track formed on a silicon wafer. The resistance of the track at $20\,°\text{C}$ is R_s.

What is the formula for the power which is dissipated in the resistive track?
Will the temperature of the silicon increase or decrease?
Will the resistance of the resistive track increase or decrease?
Will the power dissipation increase or decrease?
Is the system thermally stable or is it subject to thermal runaway?

22.4 The concentration of p-type carriers in a silicon wafer is increased by doping to a value 100 times that in intrinsic silicon. Calculate the ratio of p-type to n-type carrier concentrations in the doped semiconductor.

If the conductivity is proportional to the concentration of mobile carriers present, calculate the ratio of conductivity of the doped material to that of intrinsic material $\left(\text{conductivity} = \frac{1}{\text{resistivity}}\right)$.

22.5 Use the expression for the probability of an electron having an energy E, given on page 97, to calculate the probabilities of an electron in a crystal having energies of 2, 10, 50 and 100 times $0.025\,\text{mV}$. Compare these energies to the band gap energy for silicon.

Unit 23   Diode structure and operation

- A diode is formed from a single crystal. The doping changes from p-type to n-type within the crystal.

- Current flows through the forward biased diode when:

  - the p-type anode is at a positive voltage

  - the n-type cathode is at a negative voltage.

- Conventional current flows in the direction of the arrow of the diode symbol.

- The current-voltage characteristic for the diode is determined by the positions of the energy levels and the occupancy of the energy levels of the diode material.

- The depletion layer in a reverse biased diode is the region around the junction from which mobile charge carriers have been removed.

- The depletion layer forms a capacitance whose thickness depends on the reverse bias voltage.

- Ohmic contacts are made between the silicon crystal and the connecting wires.

- The bulk resistance associated with the silicon crystal and connections is of the order of $1\,\Omega$ and limits the maximum forward current through the diode.

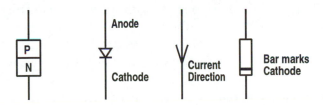

Figure 23.1: Junction diode.

A piece of p-type or n-type doped semiconductor, electrically speaking, is simply a resistor without any special properties. The valuable properties of semiconductors only appear when there is a change of doping or junction between two doped regions within a single piece of crystalline semiconductor.

We can visualize a bar of semiconductor which has been fabricated with p-type material at the top and n-type material at the bottom as shown in Figure 23.2 (a). Typical actual dimensions would be a cross section of $1\,\text{mm}^2$

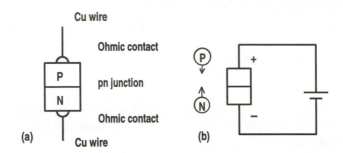

Figure 23.2: Junction diode construction and operation.

with a height of 20 microns and a junction thickness of 1 micron. The diagram in Figure 23.2 (a) has been magnified in height. Ohmic contacts are made to the top and bottom of the semiconductor crystal by bonding on contacts using metals such as gold or aluminium.

Suppose that the diode is put into a circuit as shown in Figure 23.2 (b), with a positive voltage applied to the p-type anode and a negative voltage applied to the n-type cathode. Looking in detail at the processes within the semiconductor, we can see that the p-type holes in the p-type anode region are pushed by the applied electric field down towards the junction and cross the junction into the n-type material to become minority carriers within the cathode. These holes then recombine with an n-type electron or else move to the cathode ohmic contact where they are neutralized by an incoming electron. In either case the result is simply that there is a transfer of positive charge from the anode to the cathode, in other words an electric current flows.

Within the cathode, the n-type carriers are attracted to the positive voltage on the anode, cross the junction and again contribute to the flow of charge around the circuit. So when the p-type anode is positively biased and the n-type cathode is negatively biased, current flows around the circuit. The diode is then said to be forward biased.

Now consider the case shown in Figure 23.3 where the bias voltage is reversed, with the anode negative and the cathode positive. In this case the

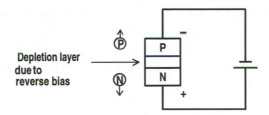

Figure 23.3: Junction diode in reverse bias.

p-type carriers in the anode are attracted away from the junction by the negative voltage on the anode. Similarly the n-type carriers in the cathode region are also attracted away from the junction by the positive voltage applied to the cathode terminal. The net effect is that the majority carriers in each region are pulled away from the junction to leave a region depleted of mobile charges called the depletion layer. There will therefore be no significant current through the diode when a reverse bias is applied. We then have the result that a diode is a device which passes electrical current in one direction but blocks the flow of current in the opposite direction.

However, even when the diode is reverse biased, a very small current flows. This reverse current arises because within the p-type anode region there are some minority carriers whose concentration is given by the semiconductor equation, $n \times p = n_i^2$. Similarly, within the n-type cathode region there are also some p-type carriers present. The reverse bias on the diode is only a reverse bias for majority carriers. For minority carriers it is a forward bias. So we get a very small constant current in a reverse biased diode which depends on the temperature and on the doping concentrations in anode and cathode regions which control the minority carrier concentrations. The reverse current is, however, independent of the reverse bias voltage. At room temperature the reverse current is typically 10^{-10} A for a silicon diode and 10^{-6} A for a germanium diode.

Silicon and germanium have different band gap energies and therefore different intrinsic carrier concentrations, n_i, so for comparable doping levels it is reasonable to expect that the minority carrier concentrations are different and therefore the reverse bias currents are also different.

One fabrication problem relates to making a connection to the semiconductor material. If a metal, which is essentially an n-type conductor, is connected to a p-type semiconductor then, in principle, a pn junction is formed which would have some rectifying properties. This problem is avoided and a fully ohmic or nonrectifying contact is formed when a metal with a work function greater than that of the p-type semiconductor work function is used to make the contact. An ohmic contact is made to an n-type semiconductor by using a metal having a work function less than that of the semiconductor work function. Choice of a different metal, with a different work function, for the contact metal can lead to the formation of a diode called a Schottky metal semiconductor junction diode.

(The work function of a metal or semiconductor is the energy difference between an electron within the crystal having an energy at the Fermi energy level and an electron moving freely outside the crystal. Typical work functions are of the order of a few electron volts.)

A historical example of such a Schottky diode is the cat's whisker diode which was used as the rectifier in the very first radio receivers and which consisted of a metal point poked at a piece of galena crystal (lead sulphide ore) until a rectifying diode was formed. A steady hand was required to maintain the rectifying contact for the duration of the radio program!

23.1 Problems

23.1 The capacitance of a parallel plate capacitor is given by $C = \frac{\kappa \epsilon_0 A}{d}$. The area A of a particular diode junction is $0.5 \, \text{mm}^2$. The capacity of the reverse biased diode is measured using a Boonton capacitance meter and is found to be $40 \, \text{pF}$. Calculate the thickness of the depletion layer, d. The dielectric constant, κ, of silicon can be taken to be 5.0. If the reverse bias is increased, will the junction capacitance increase or decrease? $\epsilon_0 = 8.854 \times 10^{-12} \, \text{F m}^{-1}$.

23.2 A varactor diode or varicap is a diode designed to have a large depletion layer capacitance when used in reverse bias. One formula for the capacitance of a varactor diode as a function of the reverse bias voltage is:

$$C(V) = \frac{C_0}{\left(1 + \frac{V}{\phi}\right)^{0.44}}$$

where $C_0 = 80 \, \text{pF}$ and $\phi = 0.6 \, \text{V}$.

If such a diode is used, connected in parallel with an inductance of $12 \, \mu\text{H}$ in the tuning circuit of a radio, plot a graph of the resonant frequency as a function of the reverse bias voltage for voltages from $0.5 \, \text{V}$ to $5.0 \, \text{V}$.

23.3 If the temperature increases, will the reverse current through a diode increase or decrease?

23.4 How do you identify which end of a diode is the cathode end?

23.5 In each of the circuits shown in Figures 23.4 (a) to (e), indicate which diodes are forward biased and which are reverse biased. Also indicate whether you expect current to flow or not to flow. Why are resistors included in the circuits?

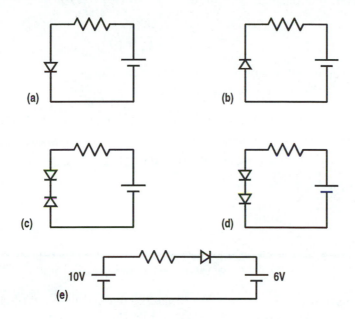

Figure 23.4: Problem 23.4.

23.6 If the resistivity of a semiconductor sample is inversely proportional to the number of mobile charge carriers present, calculate the change in the resistivity of a silicon sample when the n-type doping is increased from the intrinsic concentration to three times the intrinsic concentration.

Unit 24   Diode junction characteristics

- The current through a pn diode junction is given by:

$$I = I_0 \left(\exp\left(\frac{eV}{kT}\right) - 1 \right)$$

  where I_0 is the magnitude of the diode reverse current and V is the voltage across the diode.

- For a forward biased diode at room temperature, this reduces to:

$$I = I_0 \exp\left(\frac{V}{25\,\text{mV}}\right)$$

$$\frac{kT}{e} = 25\,\text{mV}$$

The equation $I = I_0 \left(\exp\left(\frac{eV}{kT}\right) - 1 \right)$ can be derived by a detailed analysis of the flow of charge carriers across the pn junction based on the distribution of carriers as a function of energy (Fermi-Dirac statistics). The derivation is treated in many texts on solid state physics (see, for example, *The Physics of Semiconductor Devices*, D. A. Fraser, OUP). The derivation is beyond the scope of this course and we will simply use the result.

Some points must be noted about this equation. If the voltage across the diode is zero, then $\exp(\frac{eV}{kT}) = 1$ and the junction current is zero. If the voltage across the diode is negative, corresponding to a reverse bias, then the term $\exp(\frac{eV}{kT}) \ll 1$ and the junction current is a constant $-I_0$ as expected. When a forward bias is applied, the current increases exponentially. A few representative values of junction currents, calculated using the diode current equation at room temperature, $T = 293\,\text{K}$, are shown in the table.

A cursory inspection of these calculated currents shows that the currents for forward bias voltages of 0.6 V and greater are not realistic currents. Other factors come into action to limit the currents at these forward voltages. For instance, a conductor carrying 100 A would have to have a cross section of about $100\,\text{mm}^2$ instead of the $0.5\,\text{mm}^2$ wire used in a diode if the wire is not to melt like a fuse.

Voltage V	Current A
-10	-1×10^{-10}
0	0
0.1	5.5×10^{-9}
0.4	0.9×10^{-3}
0.5	4.8×10^{-2}
0.6	2.65
0.7	144
0.9	4.3×10^{5}

It is found, however, that the currents measured for small voltages across a diode (less than 0.5 V) correspond well to these predicted currents.

A very useful simplification is made if the numerical values for Boltzmann's constant, $k = 1.38 \times 10^{-23} \, \mathrm{JK}^{-1}$, the electronic charge, $e = 1.6 \times 10^{-19} \, \mathrm{C}$ and an assumed room temperature of $t = 20 \, °\mathrm{C} = 293 \, \mathrm{K}$ are substituted to give:

$$\frac{kT}{e} = \frac{1.38 \times 10^{-23} \times 293}{1.6 \times 10^{-19}} = 0.025 \, \mathrm{V} = 25 \, \mathrm{mV}$$

The junction diode forward current then becomes:

$$I = I_0 \exp\left(\frac{V}{25 \, \mathrm{mV}}\right)$$

This is a result which we will use very frequently in our analysis of circuits.

24.1 Problems

24.1 Explain why the diode reverse current appears as a multiplying factor in the equation for the diode forward current.

24.2 If the reverse current for a particular diode is $I_0 = 5 \times 10^{-10}$ A, calculate the diode current for forward voltages of 0.25 V, 0.26 V, 0.40 V, 0.41 V. Estimate the rate of change of current with voltage, $\frac{dI}{dV}$, at 0.25 V and at 0.40 V.

24.3 Plot the log of the calculated current versus the forward voltage for a diode junction. Use the data in the table at the top of this page.

24.4 A particular diode has a reverse current, $I_0 = 10^{-10}$ A. The forward bias voltage is set at 0.45 V. Plot a graph of the diode forward current as a function of temperature for a temperature range from 10 °C to 80 °C.

Unit 25   Characteristics of real diodes

- Real diodes are modelled by an ideal junction diode in series with a bulk resistance, R_B.

Figure 25.1: Model for a real diode.

- The characteristic curve of I versus V has a knee voltage.

Material	Approximate knee voltage
Ge	0.3
Si	0.7
GaAs	2.0
SiC	2.7

- Below the knee voltage, the ideal diode equation:

$$I = I_0 \exp\left(\frac{eV}{kT}\right) = I_0 \exp\left(\frac{V}{25\,\mathrm{mV}}\right)$$

dominates the characteristic of a real diode.

- Above the knee voltage, the bulk resistance, R_B, dominates the characteristic of a real diode.

- R_B is typically of the order of $1\,\Omega$. The bulk resistance prevents the large currents which are predicted by the junction equation for large forward bias voltages from flowing in real diodes.

The measured characteristic of a silicon diode is shown in Figure 25.2.

The distinguishing feature of this actual diode characteristic is that for voltages less than 0.7 V the current is so small that it hardly shows on the graph and above 0.7 V the current-voltage relationship is linear.

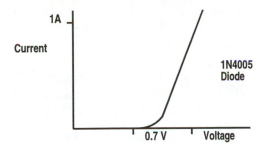

Figure 25.2: *I–V* characteristic for a 1N4005 diode.

For voltages below about 0.7 V the current is controlled by the diode junction and is related to the voltage by:

$$I = I_0 \exp\left(\frac{V}{25\,\mathrm{mV}}\right)$$

This is called the small signal region of operation.

For voltages above 0.7 V, the exponential junction relationship no longer governs the current through the diode. The current is now limited by the bulk resistance, R_B, of the diode which is due to the series resistances of the p-type and n-type material, the ohmic contacts and the connecting leads.

We can therefore model this real diode by a pn junction with the exponential characteristic and a series bulk resistance as shown in Figure 25.1.

The transition voltage between the two types of behaviour is called the knee voltage and the values of the knee voltages for different types of semi-conductor material are shown in the table in the summary.

The bulk resistance depends on the design application for the diode. A rectifier diode designed to pass currents of the order of 5 A will have a bulk resistance of the order of $R_B = 0.1\,\Omega$. A small point contact diode, which is optimized for high operating frequency but which will only pass small currents, will have a bulk resistance of the order of $R_B = 5\,\Omega$.

The bulk resistance is determined by measuring the change of current for a change of voltage in the straight section of the characteristic above the knee voltage. This is called a dynamic resistance. For an ordinary resistor the dynamic resistance and the resistance are the same but this is not the case for a diode (or any other nonlinear device).

25.1 Example

25.1 A 1N4005 type silicon diode passes 1.0 A for a forward voltage of 1.1 V.
Sketch the diode characteristic and calculate the diode bulk resistance.

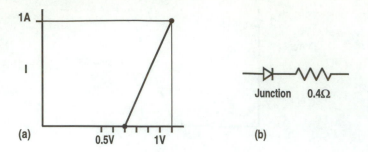

Figure 25.3: Approximation of diode characteristic.

Draw the I–V axes as shown in Figure 25.3 (a). Mark the point (1.1 V, 1.0 A). Mark the point (0.7 V, 0.0 A). Join these points as shown. This is an approximation to the diode characteristic. The real diode characteristic would be rounded off at the corner at the knee voltage.

$$\text{Bulk resistance} \quad R_B = \frac{1}{\text{slope}} = \frac{1.1\,\text{V} - 0.7\,\text{V}}{1.0\,\text{A} - 0.0\,\text{A}} = 0.4\,\Omega$$

The equivalent circuit for this diode is shown in Figure 25.3 (b).

25.2 Problems

25.1 A variable current source drives a current through a diode for which $I_0 = 10^{-10}$ and using the circuit shown in Figure 25.4. Calculate the voltages across the diode for currents of 0.1 mA and 5 mA through the diode. Assume that the diode bulk resistance is 0.5 Ω.

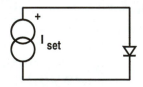

Figure 25.4: Problem 25.1.

25.2 What is the power rating for a small electronic soldering iron? Is the tip hot to touch after the iron has been switched on for five minutes?

25.3 What is the power rating for a typical torch bulb? How hot is the glass of the bulb when the bulb is lighting?

25.4 A current of 0.8 A flows through a 1N4005 diode. Calculate the power dissipation in the diode. (See Example 25.1 for data on this diode.)

25.5 The voltage across a 1N4005 diode is measured to be 0.85 V. Calculate the power dissipation in the diode. If you put your finger on the diode, should it feel hot?

25.6 The table gives spot measurements of V and I for seven different diodes. Sketch the characteristic curves for these diodes and obtain and fill in estimates for the knee voltages and for the bulk resistances of each diode type. Identify the semiconductor material used in the diode.

Diode type	Voltage	Current	Knee voltage	Bulk resistance
1N4148	1.0 V	80 mA		
	0.8 V	30 mA		
	0.5 V	50 μA		
1N4005	1 V	200 mA		
	0.8 V	100 mA		
	0.6 V	2 mA		
	0.4 V	20 μA		
OA91	0.7 V	1 mA		
	0.5 V	100 μA		
	0.4 V	2 μA		
OA47	0.6 V	50 mA		
	0.3 V	2 mA		
	0.25 V	200 μA		
	0.2 V	50 μA		
	−5.0 V	−0.1 μA		
Red led	3.0 V	60 mA		
	2.5 V	40 mA		
	2.0 V	10 mA		
	1.8 V	0 mA		
Green led	3.0 V	90 mA		
	2.5 V	45 mA		
	2.0 V	10 mA		
	1.8 V	0 mA		
Blue led	4.0 V	40 mA		
LB5410	3.0 V	10 mA		
	2.5 V	0 mA		

Unit 26   Diode circuits

When a diode is forward biased it is found in most practical cases that:

- For small currents, the voltage across the diode is the the knee voltage.

- For large currents, the voltage across the diode is given by:

$$V = V_k + I \times R_B$$

where V_k is the knee voltage as given in the summary for Unit 25.

It is found that, in the majority of applications, the current through a diode is at least 1 mA. Inspection of the diode characteristic curve in Figure 25.2 shows that the expected value for the voltage across the diode is then close to the knee voltage for currents in the range from 1 mA to about 100 mA. This is called the 'small signal region'. We therefore have the rule of thumb that, for small currents through the diode, the voltage across the diode is equal to the knee voltage.

For larger currents, such as those in rectifier circuits, we have to take into account the additional voltage drop across the diode bulk resistance of about 1 Ω. Generally speaking this correction will not amount to more than 0.4 V since diodes capable of carrying high currents (eg 10 A) are also designed to have low bulk resistances, much less than the 1 Ω of the 1N4005 type diodes.

The voltage across a forward biased diode is then given by:

$$V = V_k + I \times R_B \approx V_k \quad \text{for small currents}$$

It sometimes happens that components fail, so this rule is very useful for locating a faulty diode. If the circuit diagram indicates a forward biased diode when the power is on, then ≈0.7 V is the expected voltage across the diode. If the voltage is 0 V then the diode has failed to a short circuit or some external path is shorting the diode. If the voltage across the diode is greater than 1 V for a silicon diode then the diode is probably faulty and open circuit.

It is very easy to test a diode for diode action. Use a digital multimeter and set the range switch to the diode symbol on the resistance range. Disconnect at least one lead of the diode from the circuit. Measure the diode resistance in the forward and reverse directions. If the resistance is infinite in one direction and small in the other direction then the diode is still functioning. In most cases this is all that is required as it is unusual for a diode characteristic to deteriorate. The diode failure is usually total. It is also worth repeating your measurement for a new and unused diode to verify your procedure.

26.1 Examples

You may assume that any diode used in the examples and problems in this unit is a silicon diode having a knee voltage of 0.7 V, unless otherwise stated.

26.1 Calculate the current which flows in the circuit in Figure 26.1.

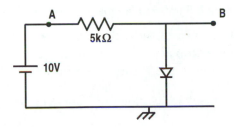

Figure 26.1: Example 26.1.

By inspection, the diode is forward biased. No large currents flow because of the 5 kΩ current limiting resistor, so we are working in the small signal region and therefore expect to have 0.7 V across the diode.

Therefore we have the equation:

$$10\,\text{V} = I \times 5\,\text{k}\Omega + 0.7\,\text{V}$$

(You should always attempt to set up an equation of this form as a starting point in the solution of electronic circuit problems.)

This equation then reduces to:

$$I = \frac{10\,\text{V} - 0.7\,\text{V}}{5\,\text{k}\Omega} = \frac{9.3\,\text{V}}{5\,\text{k}\Omega} = 1.86\,\text{mA}$$

Measuring from ground (which we always presume is the line at the bottom of the circuit diagram), the voltage at point B is +0.7 V and the voltage at point A is +10 V. The voltage from B to A is 9.3 V.

If possible, you should set up this circuit in the lab and carry out these measurements yourself.

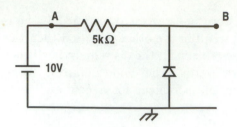

Figure 26.2: Example 26.2.

26.2 Calculate the current which flows in the circuit in Figure 26.2.

By inspection, the diode is reverse biased. This means that the current in the diode is about 10^{-10} A. The voltage drop across a 5 kΩ resistor passing 10^{-10} A is 0.5 μV which can be taken to be zero.

The voltage at one end of the 5 kΩ resistor is +10 V. Since there is no significant voltage drop across the resistor, the voltage at the other end is the same. Therefore the voltage at point B is +10 V.

26.3 Calculate the current which flows in the circuit shown in Figure 26.3.

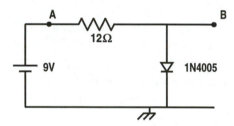

Figure 26.3: Example 26.3.

In this circuit, if the diode is replaced by a short, the current is:

$$\frac{9\,\text{V}}{12\,\Omega} \approx 0.75\,\text{A}$$

which is large, so we must use the large current approximation for the voltage across a diode:

$$V = V_k + I \times R_B$$

The required equation is then:

$$9\,\text{V} = 0.7\,\text{V} + I \times R_B + I \times 12\,\Omega$$

The diode type is a 1N4005 which we examined in Example 25.1. The bulk resistance was found to be $0.4\,\Omega$ so the equation becomes:

$$9\,\text{V} = 0.7\,\text{V} + I \times (0.4\,\Omega + 12\,\Omega)$$
$$\text{which gives}\quad 8.3\,\text{V} = I \times 12.4\,\Omega$$
$$\text{and a current}\quad I = 0.67\,\text{A}$$

26.2 Problems

26.1 Calculate the current which flows in the circuit in Figure 26.4 and also calculate the voltages at points A and B measured with respect to ground.

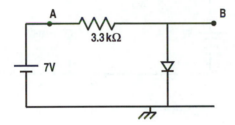

Figure 26.4: Problem 26.1.

26.2 Calculate the voltages at points A, B and C in the circuit in Figure 26.5.

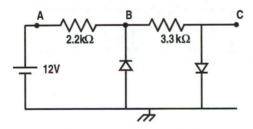

Figure 26.5: Problem 26.2.

26.3 Calculate the current which flows in the circuit in Figure 26.6 and also calculate the voltages at points A, B, C and D.

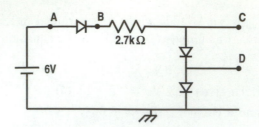

Figure 26.6: Problem 26.3.

26.4 Calculate the current which flows in the circuit in Figure 26.7 and calculate the voltages at points A and B.

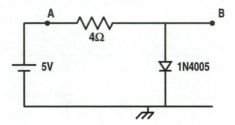

Figure 26.7: Problem 26.4.

26.5 Calculate the current which flows in each of the resistors and in the diode in the circuit in Figure 26.8. Calculate the voltages at points A and B.

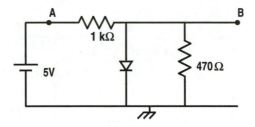

Figure 26.8: Problem 26.5.

Unit 27 Small signal diode circuits

- When a diode is operated in the small signal region, the dynamic resistance or change of voltage across the diode for a change of current through the diode is given by:

$$R_{dyn} = \frac{dV}{dI} = \frac{25\,\text{mV}}{I}$$

We have seen in Unit 24 that the current in a diode is given by:

$$I = I_0 \exp\left(\frac{V}{25\,\text{mV}}\right)$$

where V is the voltage across the diode and 25 millivolts is the calculated value of $\frac{kT}{e}$ at room temperature.

If we avoid situations where the diode current is large, then we do not need to consider the effect of the bulk resistance of the diode since it is small compared to the junction effects. This is the case when the voltage across the diode is less than the knee voltage.

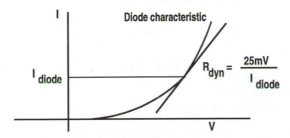

Figure 27.1: Diode characteristic and dynamic resistance.

Differentiate the diode current equation quoted above to obtain the change of current with voltage.

$$
\begin{aligned}
\frac{dI}{dV} &= \frac{d}{dV}\left(I_0 \exp\left(\frac{V}{25\,\text{mV}}\right)\right) \\
&= \frac{I_0}{25\,\text{mV}} \exp\left(\frac{V}{25\,\text{mV}}\right) \\
&= \frac{I}{25\,\text{mV}}
\end{aligned}
$$

The dynamic resistance is the reciprocal of this:

$$R_{dyn} = \frac{dV}{dI} = \frac{25\,\text{mV}}{I}$$

which, in words, is 25 millivolts divided by the DC current through the diode. This is shown in Figure 27.1

This equation for the dynamic resistance is a key tool in the analysis of diode and transistor circuits.

27.1   Examples

27.1 Calculate the change in the voltage across the diode if the setting of the power supply voltage, V_+, in the circuit in Figure 27.2, is changed from $+10\,\text{V}$ to $+11\,\text{V}$.

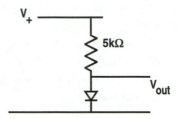

Figure 27.2: Example 27.1.

We first make the approximation that the DC voltage across the diode is $0.7\,\text{V}$.

This gives the basic equation as:

$$
\begin{aligned}
10\,\text{V} &= 0.7\,\text{V} + I \times 5\,\text{k}\Omega \\
\text{and therefore} \quad I_{10} &= \frac{10\,\text{V} - 0.7\,\text{V}}{5\,\text{k}\Omega} = 1.86\,\text{mA}
\end{aligned}
$$

For a supply voltage of $11\,\text{V}$ we get:

$$I_{11} = \frac{11\,\text{V} - 0.7\,\text{V}}{5\,\text{k}\Omega} = 2.06\,\text{mA}$$

The average DC current is $1.96\,\text{mA} \approx 2\,\text{mA}$. The dynamic resistance is then:

$$\frac{25\,\text{mV}}{I} = \frac{25\,\text{mV}}{2\,\text{mA}} = 12.5\,\Omega$$

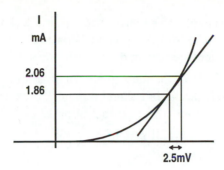

Figure 27.3: *I–V* characteristic for Example 27.1.

The change in current is $2.06 - 1.86 = 0.2\,\mathrm{mA}$. So the change in voltage across the diode is given by:

$$12.5\,\Omega \times 0.2\,\mathrm{mA} = 2.5\,\mathrm{mV}$$

This procedure is shown diagrammatically in Figure 27.3.

The alternative exact approach involves solving equations of the form:

$$V_{supply} = 5\,\mathrm{k\Omega} \times I_0 \exp\left(\frac{V_{diode}}{25\,\mathrm{mV}}\right) + V_{diode}$$

for V_{diode} with supply voltages of $10\,\mathrm{V}$ and $11\,\mathrm{V}$ and then getting the difference of the two diode voltages. You should attempt to solve this exact equation if only to prove to yourself that the linearized approach used above is significantly simpler to calculate! You could also set up this circuit in the lab and obtain some experimental values.

27.2 Calculate the output voltage waveform which would be observed on an oscilloscope connected to point A in the circuit shown in Figure 27.4.

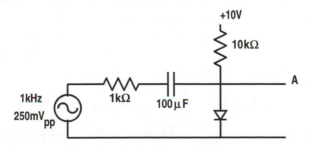

Figure 27.4: Example 27.2.

We use the principle of superposition (Unit 21) to calculate the DC component and the AC component separately and we then combine the two components at the end using the principle of superposition.

Calculate the DC component as shown in Figure 27.5 (a). We have a forward biased diode and therefore the DC voltage at the output is 0.7 V. The DC current through the diode is:

$$\frac{10\,\text{V} - 0.7\,\text{V}}{10\,\text{k}\Omega} = 0.93\,\text{mA}$$

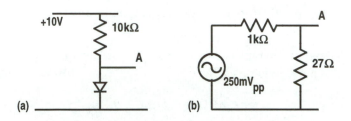

Figure 27.5: DC and AC circuits for Example 27.2.

The effective circuit for the AC component of the signal is shown in Figure 27.5 (b). The impedance of the 100 μF capacitor at 1 kHz is:

$$|Z_C| = \frac{1}{2\pi f C} = 1.6\,\Omega$$

(see Units 10 and 12). This value of 1.6 Ω is negligible compared to the 1 kΩ in series with the capacitor so the capacitor is effectively a short circuit for AC and can be omitted. (Note that the capacitor must remain in the circuit to block the DC component.) The dynamic resistance of the diode is:

$$R_{dyn} = \frac{25\,\text{mV}}{I_{DC}} = \frac{25\,\text{mV}}{0.93\,\text{mA}} = 27\,\Omega$$

The dynamic resistance of the diode and the 1 kΩ form a potential divider (Unit 4) and this is the effective circuit that is shown in Figure 27.5 (b). We can neglect the effect of the 10 kΩ which should be in parallel with the 27 Ω.

The AC component of the output is therefore:

$$V_{out\,AC} = \frac{27\,\Omega}{1\,\text{k}\Omega + 27\,\Omega} \times 250\,\text{mV}_{\text{pp}} = 6.5\,\text{mV}_{\text{pp}}$$

The signal which would be observed on an oscilloscope connected to the output at point A is a 0.7 V DC with a superimposed 6.5 mV$_{\text{pp}}$, 1 kHz AC component as is shown in Figure 27.6.

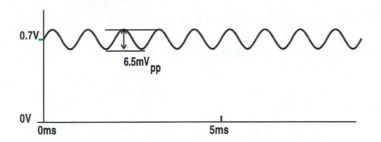

Figure 27.6: Oscilloscope output from circuit of Example 27.2.

27.2 Problems

27.1 Compare the values for the dynamic resistance calculated using the formula $R_{dyn} = \frac{25\,\text{mV}}{I}$ with the values which you estimated in Problem 24.2.

27.2 Calculate the change in the voltage at A, in Figure 27.7, resulting from a change in the voltage at B from 6 V to 7 V.

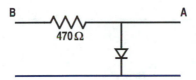

Figure 27.7: Problem 27.2.

27.3 Calculate the change in the alternating voltage at A, in Figure 27.8, when the voltage at B changes from 8 V to 12 V.

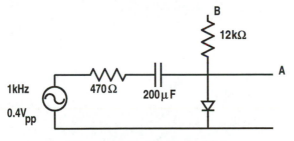

Figure 27.8: Problem 27.3.

27.4 A 20 Hz, 3 V amplitude square waveform voltage signal superimposed on a 12 V DC is applied to point B, in Figure 27.9.

A $1\,V_{pp}$ sinusoidal signal at $1\,kHz$ is applied as shown in the circuit.
Give a sketch of the input voltage waveform at B.
Give a sketch of the output voltage waveform at A.

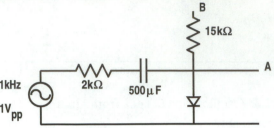

Figure 27.9: Problem 27.4.

27.5 Use the principle of superposition to calculate the voltage waveform at
the output from the circuit in Figure 27.10. Sketch the waveform which
you would observe on an oscilloscope with a time base set to $0.1\,ms$ per
division.

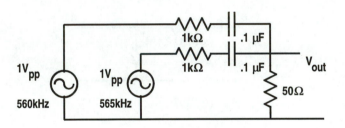

Figure 27.10: Problem 27.5.

Unit 28 Diode rectifiers

- A single diode, used as a rectifier, gives half wave rectification.

- Four diodes, in a bridge rectifier, give full wave rectification.

- Ripple voltage from a smoothed full wave rectifier $= \frac{I_{out}}{2 \times f \times C}$

When the mains 220 V_{AC} supply is used to power electronic circuits which require low voltage DC, a transformer is used to step down the AC voltage and then a single diode or a diode bridge is used to convert from AC to DC. The transformer also provides some isolation between the user and the dangerous mains voltages.

28.1 Examples

28.1 Calculate the output voltage waveform from the half wave rectifier circuit shown in Figure 28.1 (a).

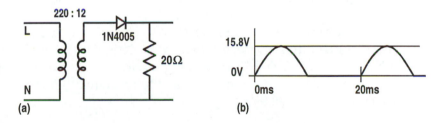

Figure 28.1: Example 28.1.

The output of this half wave rectifier circuit is across a load resistor $R_L = 20\,\Omega$ and consists of the positive half cycles of the voltage waveform out of the transformer secondary minus the voltage drop across the diode. The specification for the transformer is 220 V primary, 12 V secondary. The voltage amplitude or peak voltage (see Unit 9) at the secondary of the transformer is:

$$V_{peak} = 1.4 \times V_{RMS} = 1.4 \times 12\,V = 16.8\,V$$

121

The bulk resistance of the 1N4005 diode is $0.4\,\Omega$ so the total voltage drop across the diode is $0.7 + 0.4 \times I$. We then have the basic equation at the peak of the waveform:

$$16.8\,\text{V} = 0.7\,\text{V} + 0.4\,\Omega \times I + 20\,\Omega \times I$$

which gives $\quad I = \dfrac{16.8 - 0.7}{20.4} = 0.79\,\text{A}$

Therefore the maximum voltage across the load resistor is $20 \times 0.79 = 15.8\,\text{V}$ as shown in Figure 28.1 (b).

28.2 Give a scaled sketch of output voltage from the bridge rectifier circuit shown in Figure 28.2 (a).

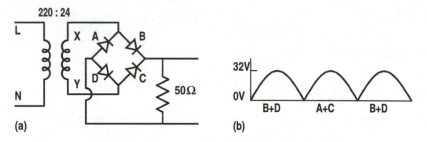

Figure 28.2: Example 28.2. Full wave rectifier circuit.

In this bridge circuit we first consider the positive half cycle of the output voltage from the transformer when X is positive with respect to Y. The diodes B and D are forward biased and the output voltage across the load resistor is positive. The effective circuit for current flow is shown in Figure 28.3 (a).

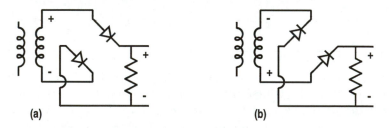

Figure 28.3: The two effective circuits for the two half cycles.

During the negative half cycle, X is negative relative to Y and diodes A and C are forward biased. This again gives a positive voltage at the top of the load resistor. The effective circuit in this case is shown in Figure 28.3 (b).

There are two diodes conducting at any one time so the basic equation for the peak voltage is:

$$24 \times 1.4 = 2 \times 0.7 + 2 \times I \times R_B + I \times 50\,\Omega$$

Taking an $R_B = 0.4\,\Omega$, this gives a peak current of $0.63\,\text{A}$ and a peak voltage across the $50\,\Omega$ of $31.7\,\text{V}$.

28.3 Calculate the output voltage from the circuit of Figure 28.4 (a).

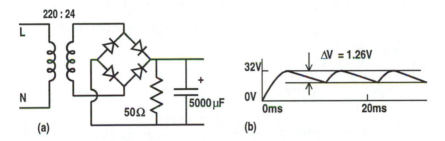

Figure 28.4: Example 28.3. Full wave rectifier with smoothing.

This is the same circuit as that in Example 28.2 but with the addition of a large smoothing capacitor of $5000\,\mu\text{F}$.

This type of large valued smoothing capacitor is manufactured using an electrolytic technology where the dielectric of aluminium oxide is formed in the wet electrolyte gel as the result of an applied voltage. The capacitor has a specific polarization and must be connected correctly into the circuit. If it is connected the wrong way around the electrolytic process goes into reverse, gas is evolved and the capacitor can explode. Always check the polarity of electrolytic capacitors before inserting them into circuits!

The output voltage waveform from this circuit is characterized by a peak which occurs at the peak of the AC waveform when the capacitor is charged up. During the dip in the rectified waveform, the capacitor acts as a reservoir and the stored charge in the capacitor maintains the current in the load for the 0.01 seconds (half the period of the 50 Hz mains) until the next peak occurs. During this time the voltage across the capacitor drops by an amount called the ripple voltage, ΔV, which is calculated as follows:

The current from the capacitor in the interval between the peaks is:

$$I_{out} = \frac{V_{out}}{R_L}$$

The charge which flows out of the capacitor is:

$$\Delta Q = \text{time} \times I_{out} = \frac{T}{2} \times I_{out} = \frac{I_{out}}{2 \times f}$$

This results in a drop, ΔV, in the output voltage which gives a ripple voltage:

$$\Delta V = \frac{\Delta Q}{C} = \frac{I_{out}}{2 \times f \times C}$$

When the component values corresponding to the circuit in Figure 28.4 (a) are inserted into this equation we get:

$$I_{out} = \frac{31.7\,\text{V}}{50\,\Omega} = 0.63\,\text{A}$$

and ripple voltage $\Delta V = \dfrac{I_{out}}{2 \times f \times C} = \dfrac{0.63}{2 \times 50 \times 5 \times 10^{-3}} = 1.26\,\text{V}$

as shown in Figure 28.4 (b).

The advantages of the full wave rectifier (which is available as a single component) are that:

- There is a more efficient use of the transformer which reduces the transformer size and weight.

- The output voltage is present for a greater fraction of the time than is the output from a half wave rectifier and therefore less smoothing is required to convert the output to a smoothed DC.

- The size of necessary smoothing capacitors is smaller and again this leads to a reduced size and weight.

28.2   Problems

28.1 Why are transformer sizes specified in V A rather than in watts?

28.2 Calculate the average output voltage and the ripple voltage for the half wave rectifier circuit shown in Figure 28.5. What will be the value of the new ripple voltage if the $470\,\Omega$ load resistor is replaced by a $10\,\mathrm{k}\Omega$ load resistor?

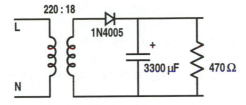

Figure 28.5: Problem 28.2.

28.3 Calculate the average output voltage and the ripple voltage for the full wave rectifier circuit shown in Figure 28.6.
Calculate the power dissipated in the load resistor.
Calculate the minimum power rating for the transformer.
What is the minimum required working voltage for the electrolytic capacitor?

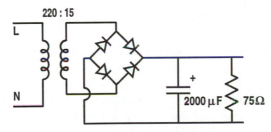

Figure 28.6: Problem 28.3.

28.4 Analyze the operation of the voltage multiplier circuit shown in Figure 28.7 by examining the flow of charge into the plates of the C_1 capacitor.

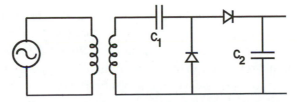

Figure 28.7: Problem 28.4.

Unit 29 Zener diodes

- Increased dopant concentrations in pn diodes reduce the reverse breakdown voltage.

- The avalanche effect dominates at large voltages. The Zener effect dominates at low voltages. The name Zener diode is used for a diode which breaks down in reverse bias due to either mechanism.

- Zener diodes are normally used in reverse bias.

- A Zener diode conducts in reverse bias when the voltage is greater than the Zener voltage for the diode.

- In a circuit the maximum temperature stability is obtained by using Zener diodes rated at about 6 volts.

Figure 29.1: Circuit symbol for a Zener diode.

If the reverse voltage across a diode is increased to large values, a voltage called the peak inverse voltage (PIV) is reached when the diode starts to conduct in the reverse direction. A typical diode such as the 1N4005 has a PIV of 600 V. This reverse conduction can cause destruction of the diode, if there is no external resistance in series with the diode to limit the current. The diode current multiplied by the voltage across the diode, $I \times V$, gives the power dissipated within the diode which causes heating and can melt and destroy the diode.

When the current is limited, removal of the large voltage allows the diode to recover fully from the breakdown. In choosing a diode for a particular application, it is important to select a diode having a PIV which is greater than the maximum reverse voltage which will ever appear across the diode in normal use and to use a safety factor of about 1.5 or more.

The voltage at which this reverse breakdown occurs can be decreased from about 1000 V in a rectifying diode down to about 3 V by increasing the dopant concentrations in the p-type and n-type regions of the diode during manufacture. The name Zener diode is used for diodes in which the reverse breakdown occurs at low voltages and Zener diodes are available having design reverse breakdown voltage or Zener voltage values extending from 2.7 V up to about 100 V. In forward bias, Zener diodes behave similarly to normal diodes. The characteristic curves of three Zener diodes are shown in Figure 29.2.

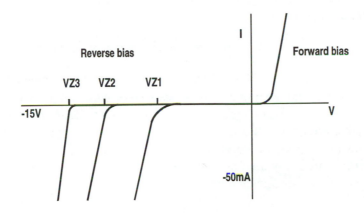

Figure 29.2: Zener diode characteristic curves.

The Zener voltage is the reverse bias voltage at which the current increases rapidly for a very small change of the reverse voltage. In the diagram $|VZ1| < |VZ2| < |VZ3|$, so the convention is to use the magnitude of the Zener voltage in specifying Zener diodes.

Two mechanisms are involved in the reverse breakdown in Zener diodes:

- For voltages greater than about 6 volts, the avalanche effect dominates. A minority carrier is accelerated across the reverse biased junction and gains enough energy to generate electron-hole pairs which themselves generate more pairs leading to a rapid increase in reverse current.

- For voltages below 6 V, the Zener effect dominates. This mechanism is due to quantum mechanical tunnelling of valence or bound electrons to nearby sites in the conduction band on the other side of the junction.

This implies that the junction must be very thin when it is reverse biased because the tunnelling current decreases exponentially with distance. The requirement of a thin junction then implies that the dopant levels in the p-type and n-type regions are very high in order to keep the depletion layer thin when the reverse bias is applied.

The name, Zener diode, is applied to a diode which is designed to exhibit breakdown at a specific voltage irrespective of which mechanism is dominant.

As the dopant concentrations in a diode are increased, the diode progresses from a normal diode with a large reverse breakdown voltage to a Zener diode with a low reverse breakdown voltage.

Two other features are worth noting in Figure 29.2.

- The sharpness of the onset of breakdown increases with increasing breakdown voltage.

- The slope of the *I–V* characteristic changes with breakdown voltage.

The temperature coefficient of the breakdown voltage is negative for the Zener mechanism and positive for the avalanche mechanism. The change over between the two mechanisms is at about 6 V at which voltage the temperature coefficient is nearly zero. If a Zener diode is to be used as a voltage reference device, the maximum temperature stability is achieved by using a Zener diode having a breakdown voltage in the region of 6 V where the temperature coefficient of Zener voltage is near zero.

29.1   Examples

29.1 The circuit shown in Figure 29.3 is a Zener diode voltage regulator. Calculate the maximum current which can be drawn by a load connected to the output before the output voltage drops below 8.2 V.

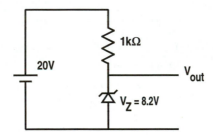

Figure 29.3: Example 29.1.

The positive voltage is applied to the cathode and therefore the diode is reverse biased and has a nominal voltage of 8.2 V across it. (Zener diodes come in the same ±10% tolerances as resistors and capacitors.) Therefore the V_{out} will be 8.2 V. The current in the 1 kΩ resistor is $\frac{20-8.2}{1\,k\Omega} = 11.8\,\text{mA}$. This 11.8 mA also flows in the Zener diode.

If some external load which draws current is connected to the output then some of this current is diverted into the external load but the output voltage remains constant at 8.2 V as long as some current flows in the reverse biased Zener diode. Therefore, as long as the output current does not exceed 11.8 mA the output voltage remains at 8.2 V. It would be better to allow a margin of safety and not allow the external load to draw more than 8 mA.

29.2 Calculate the maximum current which can be drawn from the power supply shown in Figure 29.4 without causing the output voltage to drop below 6.8 V. Calculate the minimum required power rating for the Zener diode.

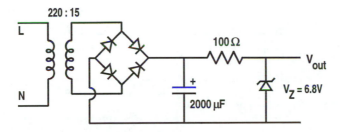

Figure 29.4: Example 29.2.

The mains supply voltage is a nominal 220 V. At times of peak load on the system the actual voltage may drop well below this value. This is sometimes called brown-out. Late at night, the voltage may increase above this nominal value. The smoothed output voltage from the rectifier will change in proportion to these mains voltage fluctuations. There will also be fluctuations due to the ripple voltage which depends on the current drawn. A resistor and Zener diode give a simple method of regulating the voltage for small output currents.

The nominal voltage across the 2000 μF capacitor is approximately 19 V (see Example 28.3) which gives a current in the 100 Ω resistor of $\frac{19-6.8}{100} = 0.12$ A. This is, in principle, the maximum regulated current available.

But the mains voltage may drop by 10% to 200 V which gives 17 V across the capacitor. Allow a 1.5 V ripple to give a worst case of 15.5 V across the capacitor. This gives a current in the 100 Ω of $\frac{15.5-6.8}{100} = 0.09$ A which is a more conservative estimate of the maximum current which can be drawn without loss of voltage regulation.

Refer back to the Zener diode characteristics in Figure 29.2. The voltage across the Zener diode changes with the current through the diode. This will also cause a change in the output voltage as the fraction of the total current flowing through the diode changes due to a change from no load current to maximum load current.

You should note that there are integrated circuit regulators available which give better stability and greater output current capacity than the simple Zener diode regulator shown here. These regulators do, however, incorporate a reference Zener in a circuit similar to the one discussed here.

29.3 Calculate the voltages at the points A, B and C in the circuit shown in Figure 29.5.

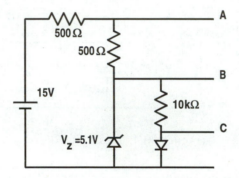

Figure 29.5: Example 29.3.

The diode is forward biased, therefore the voltage at C is 0.7 V.
The Zener diode is reverse biased, therefore the voltage at B is 5.1 V.
The two 500 Ω resistors form a potential divider between 15 V and 5.1 V.
The current in the resistors is $\frac{15-5.1}{500+500} = 9.9\,\text{mA}$.
The voltage at A is therefore $9.9\,\text{mA} \times 500\,\Omega + 5.1\,\text{V} = 10.05\,\text{V}$.
Alternatively, the voltage at A is $15\,\text{V} - 9.9\,\text{mA} \times 500\,\Omega = 10.05\,\text{V}$ as before.

29.2   Problems

29.1 Calculate the voltages at points A, B and C in the circuit shown in Figure 29.6.

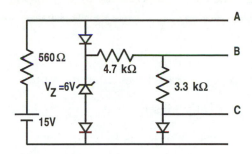

Figure 29.6: Problem 29.1.

29.2 Calculate the Thévenin equivalent for the circuit shown in Figure 29.7.

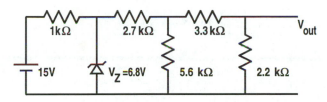

Figure 29.7: Problem 29.2.

29.3 Plot a graph of the output voltage from the circuit shown in Figure 29.8 for an input voltage which is varied from $-30\,\mathrm{V}$ to $+30\,\mathrm{V}$.

(Circuits similar to this are used in Zener barriers to limit the voltages and currents on instrumentation wires leading to hazardous areas in chemical plants where flammable gases may be present. The energy in any spark which may occur due to a fault will then be limited to an energy value below that which can cause ignition of the flammable gases which are likely to be present in the plant. Nonreplaceable fuses are also included which isolate the system in the event of a fault.)

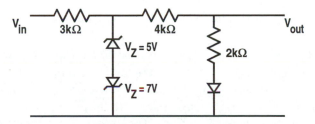

Figure 29.8: Problem 29.3.

Unit 30   Transistor structure and operation

- There are two types of transistors: npn and pnp.

- A transistor is correctly biased when:

    - The emitter-base junction is forward biased.
    - The base-collector junction is reverse biased.

- The current gain $\beta = h_{fe} = \frac{I_C}{I_B}$ and then $I_C = \beta \times I_B$

- The collector and emitter currents are approximately equal, $I_C \approx I_E$.

- Typical values for β lie in range 30 to 500.

A bipolar transistor is a three layer semiconductor device formed as either an npn or a pnp structure in a single semiconductor crystal. It is called bipolar because there are two types of semiconductor in the conduction path. The three layers are called emitter, base and collector. The fabrication methods will be discussed in more detail in Unit 36.

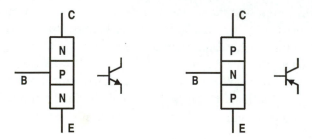

Figure 30.1: Bipolar transistor schematic structure.

The emitter region is usually more highly doped than the base or collector regions. Since there are three layers, there are two pn junctions and it is these junctions that determine the operation of the transistor. In order to bias the transistor correctly before it is used to form an amplifier, external voltages are applied so that one of these junctions, the emitter-base junction, is forward biased and the other, the base-collector junction, is reverse biased. This is shown in Figure 30.2, which is drawn for an npn transistor. For convenience, we will discuss the operation of a transistor in terms of an npn transistor but similar arguments apply to the pnp transistor, with voltages reversed and p-type and n-type materials interchanged.

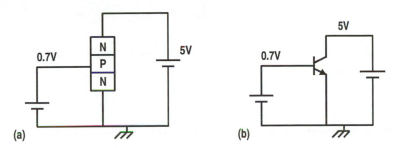

Figure 30.2: Correct biasing of npn transistor.

It is important to remember that a reverse biased pn junction is reverse biased for majority carriers only. The very small current which flows in a reverse biased pn junction diode is therefore due to the flow of minority carriers across the junction which is forward biased for minority carriers. The reverse current is so small because there are so few minority carriers present. In an npn transistor, the collector current should be very small since the base-collector junction is reverse biased and there are few electrons in the p-type base region. However, the collector current is dramatically affected by injection of electrons into the base region at the emitter-base junction.

In an npn transistor, positive bias voltage is applied between the emitter and the base to give a forward biased emitter-base diode. Since the n-type doping of the emitter region is heavier than the p-type doping of the base region, the current across the emitter-base junction will be mostly due to electrons flowing into the base region. Once within the base region, the electrons diffuse until they reach the base-collector junction. The thickness of the base region is made as small as possible to reduce the diffusion time and therefore improve the speed of the transistor. Typically the base region is 0.2 microns thick. Once the electrons, which are minority carriers in the base region, reach the base-collector junction they are swept across the junction to give a collector current.

Most of the current which enters the transistor through the emitter lead then leaves through the collector lead which gives $I_C \approx I_E$. A very small fraction, about 0.5%, of the electrons injected into the base from the emitter recombine with p-type holes in the base. This recombination would cause charging up of the base region to give an emitter-base junction which is no longer forward biased. However, a current is supplied to the base which

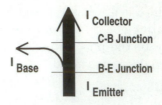

Figure 30.3: Current is proportional to line thickness.

compensates for this recombination, preventing the base charging up and maintaining the emitter-base forward bias. This base current is a nearly constant fraction of the emitter current and it is this small current into the base which controls the larger emitter current and gives the transistor its amplification properties. Figure 30.3 gives a diagrammatic representation of the various currents in the transistor.

If the current into the base of a transistor is kept constant and the collector current is measured as a function of the collector voltage we then obtain what is called the transistor output characteristic. Figure 30.4 shows the output characteristic for a typical npn small signal transistor, the BC109.

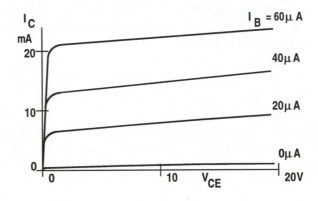

Figure 30.4: BC109 transistor output characteristic.

These characteristics would be measured using a circuit similar to that shown in Figure 30.5. The variable resistor, R_V, is varied to set the base current at the required value and the variable voltage supply is varied to sweep the V_{CE} over the range from 0 V to about 20 V.

Two very important results can be abstracted from these characteristics.

- The collector current is relatively independent of the collector voltage.

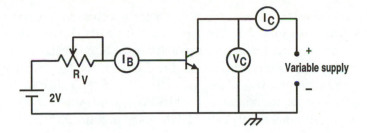

Figure 30.5: Circuit for measuring output characteristic.

- The collector current is $\beta \times I_B$ where β is the current gain for the transistor. In some data sheets you may find h_{fe} used instead of β. The symbol h_{fe} is used in the h parameter matrix method of transistor specification.

30.1   Example

30.1 Calculate the current gain at $V_{CE} = 5\,\text{V}$ for the BC109 transistor whose characteristics are shown in Figure 30.6 (a).

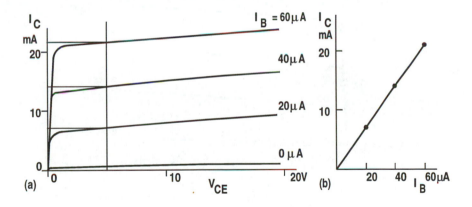

Figure 30.6: Example 30.1.

Draw a vertical line at $5\,\text{V}$. Note the collector current, I_C, for each crossing of the curves for $I_B = 0\,\mu\text{A}$, $20\,\mu\text{A}$, $40\,\mu\text{A}$, and $60\,\mu\text{A}$. Plot these values for I_C as a function of I_B as shown in Figure 30.6 (b). The slope of this curve then gives the current gain:

$$\beta = \frac{dI_C}{dI_B} \approx \frac{20\,\text{mA}}{55\,\mu\text{A}} = 363$$

It can be seen that this curve is very nearly a straight line corresponding to a constant value of β. The fact that β is nearly constant is used in the transistor parameter measuring feature, h_{fe} or β, which you will find on many digital multimeters. A fixed current of say $10\,\mu A$ is injected into the base of a transistor and the value of the collector current, I_C, is indicated on the meter but scaled as a multiple of the I_B of $10\,\mu A$. This is then numerically equal to the current gain, β, for the transistor.

30.2   Problems

30.1 Calculate the current gains for the transistor in the characteristics in Figure 30.6 (a) for collector voltages of $2\,V$ and $10\,V$.

30.2 Draw the circuit which you would use to measure the characteristics of a pnp transistor such as the BC179 (pnp version or complement of the npn BC109).

30.3 The characteristics for a 2N3055 power transistor are shown in Figure 30.7. Estimate the current gain for collector voltages of $5\,V$ and $10\,V$.

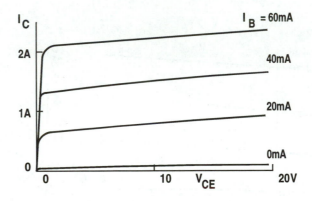

Figure 30.7: 2N3055 power transistor output characteristics.

30.4 A power transistor in a particular circuit is operated with a V_{CE} of 25 V and passes peak currents of 4.5 A. If the transistor is mounted on a heat sink which is rated at $0.9\,°CW^{-1}$, calculate the maximum temperature reached by the transistor, if the ambient temperature is 27°C.

30.5 In the event of a fault in a circuit, it is necessary to test an npn transistor to verify that it is functioning. This is normally done by testing for diode action across the emitter-base and the base-collector junctions.

Use a digital multimeter in diode test mode to measure the resistance in forward and reverse directions across the junctions.

Fill in the following table with H for high resistance and L for low resistance when the + and − terminals of the multimeter on diode test setting are connected to the indicated transistor terminals. If possible you should verify your results in the laboratory.

Would you expect a different result if a pnp transistor is tested?

+ Terminal	− Terminal	Resistance H/L
base	collector	
collector	base	
base	emitter	
emitter	base	
collector	emitter	
emitter	collector	

30.6 Figure 30.8 shows a range of transistor package types. Refer to an electronic components catalog and identify the different package type code numbers and also identify the emitter, base and collector leads or connections for each transistor package type.

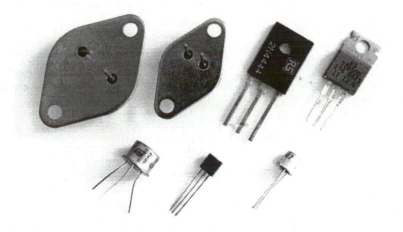

Figure 30.8: Typical transistor packages.

Unit 31 Transistor bias circuits

- The function of a transistor bias circuit is to maintain a forward bias on the emitter-base junction and a reverse bias on the base-collector junction.

- The currents and voltages in transistor bias circuits can usually be determined by using:

$$I_C = \beta \times I_B$$
$$I_C \approx I_E$$
$$V_{BE} \approx 0.7\,\text{V}$$

  to solve a basic equation which represents the sum of the voltage drops across the components along a path between the supply voltage rail and the ground rail.

- The basic equation for calculating the currents in transistor bias circuits is determined along a path such that:

  Voltage supply = Sum of voltage drops across individual components.

Before a transistor can be made to do anything useful it must be biased so that the emitter-base junction is forward biased and the base-collector junction is reverse biased. There are only a small number of circuits which are used to bias transistors and place the transistor in the middle of its operating range.

The first operation to be carried out in determining the currents and voltages is to identify a path through the circuit from the supply voltage rail to the ground rail for which the sum of the voltage drops across the individual components can be explicitly stated. We call this the basic equation. The main aim of this unit is to learn how to identify the basic equation for any problem rather than to try to remember all possible basic equations for all possible circuits.

In Examples 31.1 and 31.2, the path along which the basic equation is set up is identified in a separate sketch. In general, the basic equation will not be set up along a path which includes the base-collector junction because the current through this junction is nearly independent of the voltage across the junction and is not readily calculated.

31.1 Examples

31.1 Calculate the emitter, base and collector voltages and currents for the circuit shown in Figure 31.1 (a). Use $\beta = 250$.

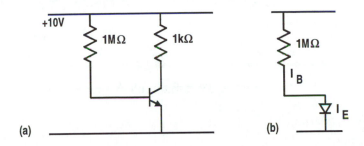

Figure 31.1: Example 31.1.

By inspection it can be seen that:

$$V_E = 0\,\text{V}$$
$$V_B = 0.7\,\text{V}$$

The path along which the basic equation is evaluated is shown in Figure 31.1 (b) and the basic equation is:

$$10\,\text{V} = I_B \times R_B + 0.7\,\text{V}$$

which gives
$$10\,\text{V} = I_B \times 10^6 + 0.7\,\text{V}$$

and then
$$I_B = \frac{10 - 0.7}{10^6}$$
$$= 9.3\,\mu\text{A}$$

$$I_C = \beta \times I_B$$
$$= 250 \times 9.3\,\mu\text{A}$$
$$= 2.3\,\text{mA}$$

This gives
$$V_C = 10\,\text{V} - I_C \times R_C$$
$$= 10\,\text{V} - 2.3\,\text{mA} \times 1\,\text{k}\Omega$$
$$= 7.7\,\text{V}$$

Note that the collector voltage, V_C, is calculated as $I_C \times R_C$ down from the supply voltage of $10\,\text{V}$.

31.2 Calculate the emitter, base and collector voltages and currents for the circuit shown in Figure 31.2 (a). Use $\beta = 300$.

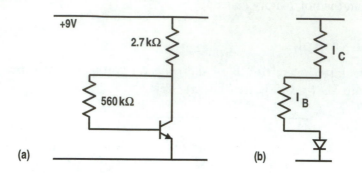

Figure 31.2: Example 31.2.

By inspection, we find that:

$$
\begin{aligned}
V_E &= 0\,\text{V} \\
V_B &= 0.7\,\text{V}
\end{aligned}
$$

From Figure 31.2 (b), the basic equation is:

$$
\begin{aligned}
9\,\text{V} &= I_C \times R_C + I_B \times R_B + 0.7\,\text{V} \\
\text{Use} \quad I_C &= \beta \times I_B \\
\text{to get} \quad 9\,\text{V} &= \beta \times I_B \times R_C + I_B \times R_B + 0.7\,\text{V} \\
\text{Solve to get} \quad I_B &= \frac{9 - 0.7}{\beta \times R_C + R_B} \\
&= \frac{8.3}{300 \times 2.7 \times 10^3 + 560 \times 10^3} \\
&= 6.06\,\mu\text{A} \\
\text{Then} \quad I_C &= \beta \times I_B \\
&= 1.82\,\text{mA} \\
V_C &= 9\,\text{V} - 1.82\,\text{mA} \times 2.7\,\text{k}\Omega \\
&= 9\,\text{V} - 4.91\,\text{V} \\
&= 4.09\,\text{V}
\end{aligned}
$$

31.3 Calculate the emitter, base and collector voltages and currents for the circuit shown in Figure 31.3.

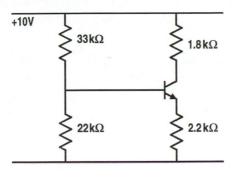

Figure 31.3: Example 31.3.

In this circuit, the basic equation can be set up along a path through the potential divider, subject to the condition that the current flowing into the base is less than 10% of the current in the resistors in the potential divider.

$$\text{Current in potential divider} \quad = \quad \frac{10\,\text{V}}{33\,\text{k}\Omega + 22\,\text{k}\Omega}$$
$$= \quad 182\,\mu\text{A}$$

which satisfies the 10% condition since the base current is usually $\approx$ $10\,\mu\text{A}$. The base voltage is then determined by the potential divider as:

$$V_B = \frac{22}{33+22} \times 10\,\text{V} = 182\,\mu\text{A} \times 22\,\text{k}\Omega = 4\,\text{V}$$

The emitter-base voltage is 0.7 V and therefore:

$$\begin{aligned} V_E &= V_B - 0.7\,\text{V} \\ &= 4.0\,\text{V} - 0.7\,\text{V} = 3.3\,\text{V} \\ \text{Then} \quad I_E &= \frac{3.3\,\text{V}}{2.2\,\text{k}\Omega} = 1.5\,\text{mA} \\ \text{and} \quad I_C &\approx 1.5\,\text{mA} \\ V_C &= 10\,\text{V} - 1.8\,\text{k}\Omega \times 1.5\,\text{mA} \\ &= 10\,\text{V} - 2.7\,\text{V} = 7.3\,\text{V} \end{aligned}$$

The interesting feature of this circuit is that the currents and voltages have been determined without having to use any specific value of the current gain, β, in the calculation. When transistors are manufactured, the value of the β has great variation from one batch to the next. From the catalogs it can be seen that the manufacturers only specify that the current gain, β, or h_{fe} for a BC109 transistor (a typical transistor) lies in the range from 200 to 800.

This type of bias circuit is usually chosen by designers and manufacturers of electronic circuits because it only requires that the value for β exceed some minimum value. The resistors then determine the operating values for the transistor. Compare this situation with Examples 31.1 and 31.2 where any variation in the value of β has a large effect on the operating point.

31.2   Problems

31.1 Calculate the emitter, base and collector voltages and currents for the circuit shown in Figure 31.4. Use $\beta = 150$.

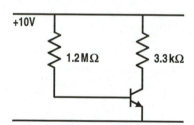

Figure 31.4: Problem 31.1.

31.2 Calculate the emitter, base and collector voltages and currents for the circuit shown in Figure 31.5. Use $\beta = 250$.

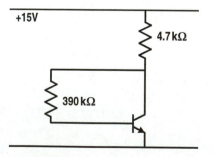

Figure 31.5: Problem 31.2.

31.3 The circuit shown in Figure 31.6 was constructed and the voltage at the collector was measured to be +6.2 V. Calculate the emitter, base and collector currents and calculate the current gain, β, for the transistor used in the circuit.

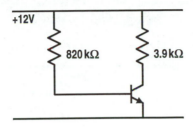

Figure 31.6: Problem 31.3.

31.4 Calculate the emitter, base and collector voltages and currents for the circuit shown in Figure 31.7. Use $\beta = 300$.

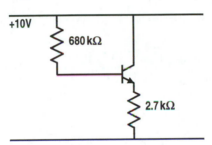

Figure 31.7: Problem 31.4.

31.5 Calculate the emitter, base and collector voltages and currents for the circuit shown in Figure 31.8. Use $\beta = 200$.

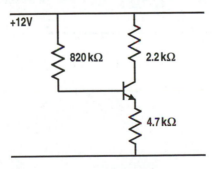

Figure 31.8: Problem 31.5.

Learning Resources
Centre

31.6 Calculate the new values of the voltages and currents in the circuit of Problem 31.5 which result from replacing the transistor by another transistor having a $\beta = 300$.

31.7 Calculate the emitter, base and collector voltages and currents for the circuit shown in Figure 31.9. The current gain for the pnp transistor is $\beta = 185$.

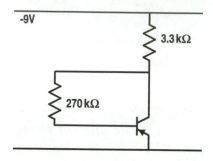

Figure 31.9: Problem 31.7.

31.8 Calculate the emitter, base and collector voltages and currents for the circuit shown in Figure 31.10.

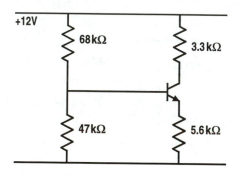

Figure 31.10: Problem 31.8.

Unit 32   Small signal amplifiers

- Lower case v or i is used to represent small signals or variations about a mean DC operating point of value V or I.

- A small signal applied to a transistor causes the instantaneous operating point to move along a load line.

- The amplification is represented by A. A negative value for A implies that the output signal waveform is inverted.

- In a common emitter amplifier, the emitter is common to the input and the output ports for small signals.

- In a common base amplifier, the base is common to the input and the output ports for small signals.

- In a common collector amplifier, the collector is common to the input and the output ports for small signals.

The discussion of transistor circuits in earlier units has only covered the DC biasing of transistors into the middle of their working range so that the emitter-base junction is forward biased and the base-collector junction is reverse biased. No signal has, as yet, been applied to the transistor.

In this unit we examine ways of getting some useful signal gain from the transistor circuit. We first define a small signal as a fluctuation in voltage or current about an operating or bias point and we will use lower case i or v, with appropriate subscripts for emitter, base, collector etc., to represent these small signals. These small signals are superimposed on the DC bias voltages and currents. They modulate the currents and voltages in the transistor and can be amplified by the action of the transistor.

Consider the circuit in Figure 32.1 (a) which has already been discussed in Example 31.1. The collector voltage and current were calculated to be $V_C = 7.7\,\text{V}$ and $I_C = 2.3\,\text{mA}$ and this (I_C, V_C) pair is indicated as the operating point on the load line in Figure 32.1 (b).

145

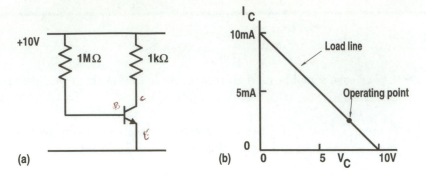

Figure 32.1: Transistor load line.

Suppose we could change or modulate the base current in some way. The relationship between the resulting collector current and collector voltage would still be:

$$V_C = V_{supply} - I_C R_C = 10\,\text{V} - I_C \times 1\,\text{k}\Omega$$

which is a straight line. This straight line is easily drawn on the I_C–V_C diagram. When $I_C = 0$, $V_C = 10\,\text{V}$ and for a short circuit across the transistor $V_C = 0$ and $I_C = \frac{10\,V}{R_C} = 10\,\text{mA}$. We can therefore draw the load line on the V_C–I_C diagram as shown in Figure 32.1 (b). The calculated operating point also lies on this load line. Any variation in the signal to the base which we can impose only causes the (V_C, I_C) value to move up or down along this load line.

Our requirement is that we can bias the transistor so that it is at an operating point towards the middle of this load line and that we have a method of superimposing some small variation on the base current or voltage of the transistor which will cause the operating point to vary along the load line without disturbing the mean DC bias values.

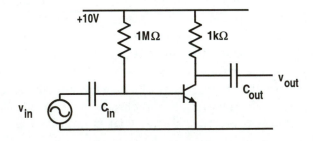

Figure 32.2: Method of coupling in small signals.

This can be achieved by using the circuit shown in Figure 32.2 where we superimpose an input signal, v_{in}, from a function generator onto the DC bias signal, V_B, at the base by using a capacitor to couple in the signal. The capacitor prevents the function generator from shorting out the DC bias voltage but the capacitor also acts as a low value impedance for the small input signal from the function generator, v_{in}, allowing the signal to be superimposed on the DC base bias. (Review Example 27.2, which shows a similar example of superposition for a diode circuit.)

Without going into detailed calculations of the magnitudes of the signals, it can be seen that the sequence of changes is:

- An increased voltage at the base causes an increased base current.

- The increased base current is amplified by the transistor with a current gain, β, and gives a larger emitter current.

- The larger emitter current gives a nearly equal change in the collector current.

- The increased collector current gives an increased voltage drop across the collector resistor, R_C.

- The increased voltage drop across the collector resistor gives a decrease of the collector voltage.

- This decrease in the collector voltage implies an inversion of the signal.

If a capacitor is connected to the collector as C_{out} in Figure 32.2, then the DC component of the output voltage can be removed to leave only the time varying small signal output voltage, v_{out}.

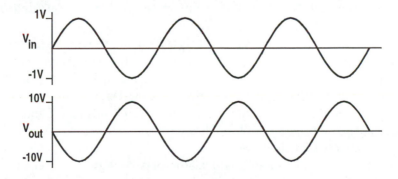

Figure 32.3: Oscilloscope trace of input and output signals for $A = -10$.

If the values of the resistors are chosen correctly then the output signal, v_{out}, may be an inverted and amplified version of the input signal, v_{in}, and we can then define a small signal amplification:

$$A = \frac{v_{out}}{v_{in}}$$

A negative value for A implies signal inversion, that is a positive going input signal gives a negative going output signal and vice versa. The magnitude of A gives the ratio of the amplitudes of the two signals. Figure 32.3 shows the signals which would be observed with an oscilloscope connected to an amplifier for which the gain is $A = -10$. Note the scaling of the voltage axes and also the inverted output waveform.

In an amplifier, the input signal, v_{in}, is applied via two wires and the output signal, v_{out}, is taken out via two wires. We can then say that the amplifier is a two port device where the input port has two connections and the output port has two connections. This is shown in Figure 32.4 (a).

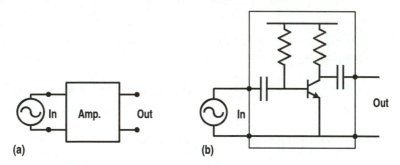

Figure 32.4: Two port amplifier.

However, transistors have only three connections (E, B and C) so one of the transistor connections must be common to both the input and output ports as is shown for the amplifier in Figure 32.4 (b). In this case, the common transistor terminal is the emitter and this type of amplifier is therefore called a common emitter amplifier. The emitter need only be common to the input and output for small signals. In this circuit example the emitter is actually connected to both the input and output. However, in some circuits the common terminal is only common to the input and output for small signals with the connection being made with a capacitor, either directly to ground or via the capacitor in the power supply.

If we take the most general transistor bias configuration, such as that discussed in Example 31.3, then the three possible amplifier configurations are as shown in Figure 32.5.

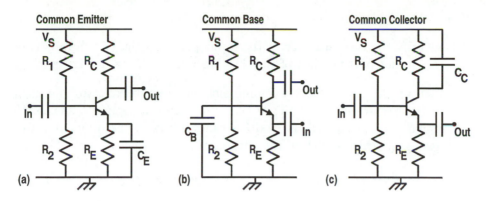

Figure 32.5: Common emitter, base and collector configurations.

For the common emitter circuit in Figure 32.5 (a) the emitter is grounded via the C_E so the emitter voltage does not vary in the presence of small signals at the input. The R_E and C_E act as a high pass filter with a corner frequency lower than the lowest signal frequency which is to be amplified in the circuit. The input signal is applied to the base and the amplified signal is taken from the collector.

For the common base amplifier in Figure 32.5 (b), the base is grounded for small signals via the C_B. Note that the C_B does not disable the DC base bias of the transistor. The input signal is applied at the emitter. A positive going input signal causes the V_{BE} to reduce thus reducing the emitter and collector currents in the transistor. This gives an increase in the collector voltage and therefore the sign of the amplification, A, is positive denoting no inversion. This type of common base amplifier has applications in high frequency and in radio frequency amplifiers.

For the common collector amplifier in Figure 32.5 (c) the collector is grounded via the C_C and the capacitor in the output of the power supply. A positive signal applied to the base causes the base current to increase and this causes the emitter current to increase. This gives increased voltage across the R_E and therefore the emitter voltage follows the base voltage. This type of amplifier is sometimes called an emitter follower. A consequence of this emitter following action is that the voltage amplification of this circuit is +1 or slightly less than +1. The current gain is, however, quite high and also the input impedance of the circuit is high. The applications of this circuit are as input stages where high input impedances are required and as output stages where the low output impedance or the capability for driving high currents into loads is needed.

32.1   Problems

32.1 Identify the two terminals of the input port and the two terminals of
the output port for each of the three amplifiers shown in Figure 32.5.

32.2 Identify the circuit type (common emitter, base or collector) for each
of the circuits shown in Figure 32.6. In each case, apply a sinusoidal
signal of amplitude 10 mV and frequency 5 kHz to the input and give
a roughly scaled sketch of the output voltage waveform.

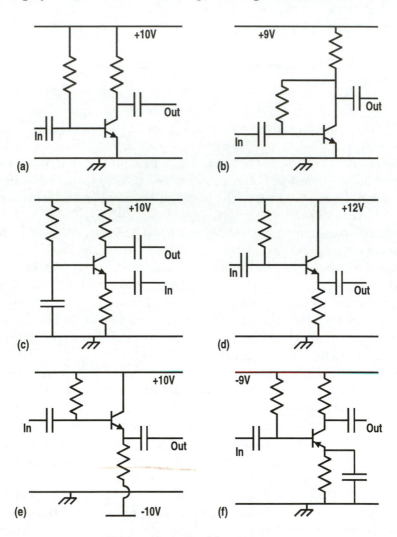

Figure 32.6: Problem 32.2.

32.3 Plot the load line for the circuit shown in Figure 32.7 and indicate the position of the operating point on the load line.

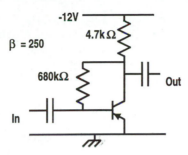

Figure 32.7: Problem 32.3.

32.4 Identify the type of circuit shown in Figure 32.8 and describe the operation of the circuit and the circuit characteristics.

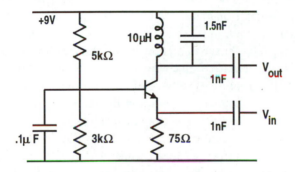

Figure 32.8: Problem 32.4.

32.5 In the circuit shown in Figure 32.9, V_{in} is varied from $0\,\text{V}$ to $+10\,\text{V}$. Plot a graph showing the variation of V_{out} as a function of V_{in}.

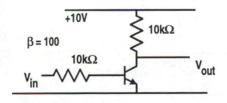

Figure 32.9: Problem 32.5.

Unit 33 Small signal amplification

- The small signal voltage amplification of a common emitter amplifier is given by:

$$A_V = -\frac{I_E}{25\,\text{mV}} \times R_C$$

- The input impedance is given by:

$$R_{in} = \beta \times \frac{25\,\text{mV}}{I_E}$$

In the last unit we saw how various configurations of transistor and resistors can give amplification of small signals which are applied to the input port of the amplifier. In this unit we show how the numerical value of the voltage amplification, A_V, can be calculated for the case of the common emitter amplifier. It is important to distinguish between the current gain, β, of the transistor and the amplification, A_V, of the complete transistor circuit. The amplification of the circuit is determined by the value of β for the transistor in conjunction with the values of the resistors used in the circuit.

Consider the stripped down circuit shown in Figure 33.1 (a). We presume that the transistor is suitably biased, that is that the emitter-base junction is forward biased and that the base-collector junction is reverse biased. For clarity we have omitted the bias resistors from the diagram.

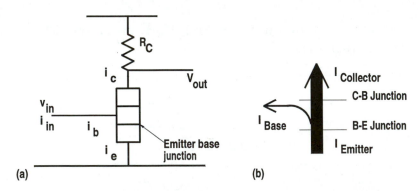

Figure 33.1: Relative magnitudes of currents in a transistor.

Consider only the small signals which we represent by the lower case letters, v and i. The small input voltage signal, v_{in}, and the resulting current signal, i_{in}, are applied to the circuit at the input port. We can then define an input resistance for small signals:

$$R_{in} = \frac{v_{in}}{i_{in}}$$

We also have the equalities $v_b = v_{in}$ and $i_b = i_{in}$.

It is assumed that the input signals are small enough that any deviation from a linear approximation is negligible. This can be very easily checked in the final circuit by using an oscilloscope to observe the output waveform. If the output waveform changes only in amplitude as the input signal amplitude is increased then the system is effectively a linear system. If the shape of the output waveform becomes distorted then the system is nonlinear.

The change in the voltage at the base appears as a change in the voltage across the emitter-base junction. This gives a change in the current through the emitter-base junction of i_e. In words, a change in the **base** voltage gives a change in the **emitter** current which flows through the emitter-base junction. Most of this emitter current flows on through the base region to the collector with only a small fraction appearing as i_b. This is indicated by the line thickness in the diagram in Figure 33.1 (b).

The emitter-base junction has a dynamic resistance which is given by the equation:

$$R_{DYN} = \frac{25\,\text{mV}}{\text{Junction current}} = \frac{25\,\text{mV}}{I_E}$$

Note that we use the DC value, I_E, of the emitter current through the junction in this equation. (Review Unit 27.) So we now have:

$$\frac{v_b}{i_e} = R_{DYN} = \frac{25\,\text{mV}}{I_E}$$

$$\text{which gives} \quad v_{in} = v_b = i_e \times \frac{25\,\text{mV}}{I_E}$$

but we also have an output voltage signal:

$$v_{out} = -i_c \times R_C \approx -i_e \times R_C$$

Now we can get the small signal voltage amplification for the common emitter amplifier circuit as:

$$A_V = \frac{v_{out}}{v_{in}} = -\frac{i_e \times R_C}{i_e \times \frac{25\,\text{mV}}{I_E}} = -\frac{I_E}{25\,\text{mV}} \times R_C$$

The negative sign indicates that the signal is inverted. The unexpected feature of this result is that it does not contain any explicit reference to the current gain, β, of the transistor.

Now consider the input resistance. We have:

$$
\begin{aligned}
R_{in} &= \frac{v_{in}}{i_{in}} = \frac{v_b}{i_b} \\
&= \frac{i_e \times \frac{25\,\mathrm{mV}}{I_E}}{i_b} \\
&= \frac{\beta \times i_b \times \frac{25\,\mathrm{mV}}{I_E}}{i_b} \\
&= \beta \times \frac{25\,\mathrm{mV}}{I_E}
\end{aligned}
$$

So the input resistance is β times the dynamic resistance of the emitter-base junction.

33.1   Example

33.1 Calculate the amplification and the input resistance of the circuit shown in Figure 33.2. Use $\beta = 200$.

Give a sketch of the signals which would be observed on an oscilloscope connected in turn to each of the nodes of the circuit.

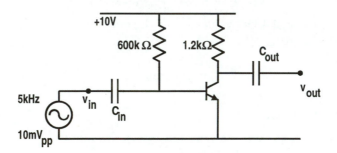

Figure 33.2: Example 33.1.

It is necessary to determine the DC operating conditions for the circuit.

$$
\begin{aligned}
V_E &= 0\,\mathrm{V} \\
V_B &= 0.7\,\mathrm{V}
\end{aligned}
$$

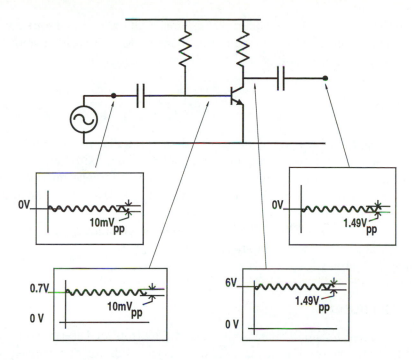

Figure 33.3: The signals at various points in a common emitter amplifier.

$$10\,\text{V} = I_B \times 600\,\text{k}\Omega + 0.7\,\text{V}$$
$$I_B = \frac{9.3}{600\,\text{k}\Omega} = 15.5\,\mu\text{A}$$
$$I_E = I_C = \beta \times I_B = 3.1\,\text{mA}$$
$$V_C = 10\,\text{V} - I_C \times R_C = 10\,\text{V} - 3.1\,\text{mA} \times 1.2\,\text{k}\Omega$$
$$= 6.28\,\text{V}$$

Now calculate the small signal voltage amplification.

$$A_V = -R_C \times \frac{I_E}{25\,\text{mV}} = -1200 \times \frac{3.1\,\text{mA}}{25\,\text{mV}}$$
$$= -149$$
$$R_{in} = \beta \times \frac{25\,\text{mV}}{I_E} = 200 \times \frac{25\,\text{mV}}{3.1\,\text{mA}}$$
$$= 1.6\,\text{k}\Omega$$

If the signal frequency is high enough for the impedance of the C_{in} to be negligible compared with $1.6\,\text{k}\Omega$ then the small signal input impedance is resistive and $1.6\,\text{k}\Omega$. Otherwise we have an effective high pass filter at the input to the amplifier which blocks DC and low frequency AC signals from reaching the transistor. This gives an amplifier frequency response (review Unit 16) with a corner frequency at:

$$f_c = \frac{1}{2\pi C_{in} R_{in}}$$

The signals which would be observed at the nodes of the circuit with an oscilloscope are shown in Figure 33.3. The solid horizontal line in the oscilloscope tracings indicates the 0 V reference on the oscilloscope screen. The peak-to-peak amplitudes of the input and output signals are indicated. If possible you should try out this circuit in the laboratory and verify these oscilloscope observations. Note the phase inversion between the input and output signals corresponding to the negative sign in $A_V = -149$.

33.2   Problems

33.1 Calculate the small signal voltage amplification for the transistor amplifier shown in Figure 33.4. The current gain for the transistor is $\beta = 250$. Plot the load line for the transistor and mark the DC operating point. Indicate the maximum and minimum excursion along the load line for a sinusoidal input signal of $12\,\mathrm{mV_{pp}}$ and of frequency $5\,\mathrm{kHz}$.

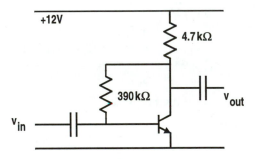

Figure 33.4: Problem 33.1.

33.2 Calculate the small signal amplification for the circuit shown in Figure 33.5. Give scaled sketches of the voltage waveforms which you would observe with an oscilloscope connected to points A, B, C and D in the circuit when a $10\,\mathrm{mV_{pp}}$, $5\,\mathrm{kHz}$ sinusoidal signal is applied at the input.

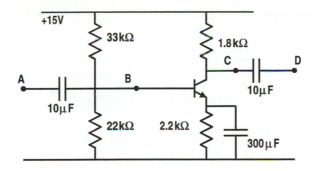

Figure 33.5: Problem 33.2.

33.3 Analyze the circuit shown in Figure 33.6 and derive the equation for the small signal voltage amplification of the circuit. You should follow the procedure used in the text but note that R_E is in series with the emitter-base junction. What is the amplification when $R_E = 25\,\Omega$?

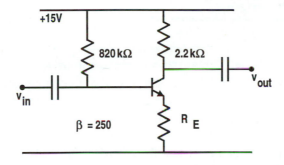

Figure 33.6: Problem 33.3.

Learning Resources
Centre

Unit 34   Transistor circuit building blocks

The characteristic features of the principal transistor circuit building blocks are:

- **Common emitter amplifier:** $A_V \approx -200$, $R_{in} \approx 3\,\mathrm{k\Omega}$

- **Emitter follower:** $A_V \approx 1$, R_{in} is high, R_{out} is low.

- **Push-pull emitter follower:** $A_V \approx 1$, high output current drive capability, symmetrical performance.

- **Differential amplifier:** Amplifies DC signals.

- **Current mirror:** Nearly constant current, used as high resistance load resistor or active load.

- **Tuned amplifier:** Frequency at which maximum amplification occurs determined by $f = \frac{1}{2\pi\sqrt{LC}}$.

In this unit we will give a brief treatment of the principal transistor circuit building blocks which are used in linear analog circuit systems. These circuit blocks or modules constitute a minimum set of such blocks that can be combined to form more complex circuits which have increased functionality.

Common emitter amplifier. We have already analyzed the performance of the common emitter amplifier in Unit 33 so we will not discuss the circuit further but just include the result in the summary.

Emitter follower. The emitter follower or common collector is shown in Figure 34.1. A signal applied at the base changes the base voltage. The emitter voltage follows the base voltage up or down maintaining a nearly constant emitter-base voltage of 0.7 V. The result is that the voltage gain is 1 or slightly less than 1. (A good analogy here is that of a tow truck and trailer. The trailer does not normally overtake the tow truck!)

The input impedance of the emitter follower is calculated as follows. A small input voltage v_{in} causes a base voltage v_b which gives base current i_b. The current across the emitter-base junction and also in R_E is $i_e = \beta i_b$. This current through the emitter-base junction and R_E in series gives a voltage at the base of:

$$v_b = i_e \left(\frac{25\,\mathrm{mV}}{I_E} + R_E \right) = \beta i_b \left(\frac{25\,\mathrm{mV}}{I_E} + R_E \right)$$

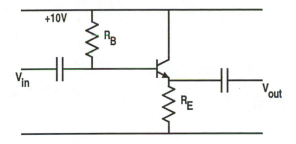

Figure 34.1: Emitter follower.

$$\text{giving} \quad R_{in} = \frac{v_{in}}{i_{in}} = \frac{v_b}{i_b} = \beta\left(\frac{25\,\text{mV}}{I_E} + R_E\right)$$

typically $R_E \approx 1\,\text{k}\Omega \gg \frac{25\,\text{mV}}{I_E}$ and we can see that the input resistance of the emitter follower will easily be of the order of $250 \times 1\,\text{k}\Omega = 250\,\text{k}\Omega$.

This makes the circuit ideal for use as the input stage in situations where we require a high input impedance to minimize the loading on sensors or other electronic pick-up devices.

The emitter follower is also used for output stages where the high current driving capability of the circuit—its current amplification properties—makes it ideal for driving low impedance loads such as loudspeakers ($R_{in} = 8\,\Omega$ typically), small, speed controlled motors and other devices which draw large currents at low voltages.

Push-pull emitter follower. The main problem with the single transistor npn emitter follower is that while the npn responds well to positive going signals, fast, large, negative going signals can give a reverse biased emitter-base junction and a turned off transistor. This occurs especially in situations where there is some capacity associated with the emitter circuit which tends to maintain the emitter voltage constant. There is therefore a good fast response for positive going signals but a poor response for negative going signals. The reverse is true for the pnp version of the emitter follower.

The solution to this asymmetric response problem is to use two transistors, one pnp and the other npn, in series as shown in Figure 34.2 (a). This gives fast response to fast input signals of either polarity but, as is normal, the solution to one problem causes another problem. The circuit in Figure 34.2 (a) has a dead spot in the response, for signals within $\pm 0.7\,\text{V}$ of zero, due to the fact that neither transistor is conducting in this region. This is called crossover distortion and Figure 34.3 shows how such a system distorts a sinusoidal input waveform.

This crossover distortion can be minimized or eliminated by providing some DC forward bias for the transistors even when there is no input signal.

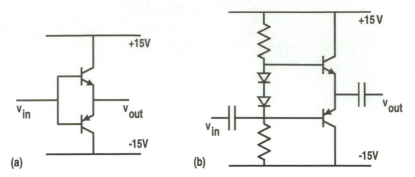

Figure 34.2: Push-pull emitter follower.

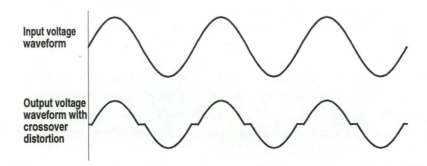

Figure 34.3: Crossover distortion.

The necessary modification is shown in Figure 34.2 (b) where the two diodes are forward biased by the resistor chain and the emitter-base junctions are maintained at $V_{BE} \approx 0.7\,\text{V}$ for each transistor. So we no longer have the problem of both emitter followers being off for small signals and therefore the crossover distortion is eliminated. Diodes are often used to give the 1.4 V between the bases rather than another resistor as the use of two diodes maintains a constant 1.4 V even if different + and − supply voltages are used. This method of minimizing crossover distortion is at the expense of increased quiescent current in the two transistors which can then be a problem in battery powered equipment.

Differential amplifier. Small AC signals can be coupled into a single transistor amplifier through a capacitor without upsetting the DC bias in the transistor. It is not possible to couple in DC signals without affecting the V_{BE} of 0.7 V and therefore it is not possible to construct a single transistor

DC amplifier which operates down to zero volts.

The method used to obtain DC amplification is to use two transistors, balanced against each other, in what is called a differential amplifier configuration. A typical circuit is shown in Figure 34.4 in which representative values are shown for the resistors.

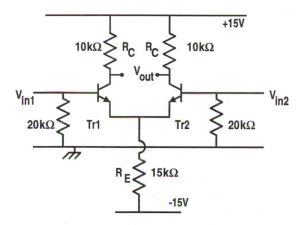

Figure 34.4: Differential amplifier.

The circuit is symmetrical, with a signal applied to the base of each transistor. The essential principle of operation is that the two halves of the circuit are coupled together by the shared emitter resistor, R_E, of 15 kΩ.

When no input signals are present at the inputs, the current in R_E divides equally between the two transistors giving equal voltages at the two collectors and therefore giving a differential output voltage between the two collectors of 0 V.

If a small positive voltage signal is applied to the base of the left hand transistor, Tr1, then this transistor conducts more current. This causes the current in R_E to increase slightly due to emitter follower action and the V_{BE} for the right hand transistor, Tr2, is reduced slightly reducing the current in Tr2. The net effect is that the voltage at the collector of Tr1 decreases and the voltage at the collector of Tr2 increases giving a change in the voltage difference between the two collectors.. The current in R_E remains nearly constant since an increase of current through Tr1 is balanced by a decrease of current through Tr2.

If the signal already being applied to the base of Tr1 is also applied to the base of Tr2 then the balance of the circuit is restored and the voltage difference between the collectors is restored to zero.

We now obtain an expression for the gain of the differential amplifier.

First set up the basic equation (see Unit 31) by calculating the voltage drops along the path from the ground line to the base of Tr1, through the emitter-base junction of Tr1 and then through the R_E to the -15 V supply. We then get the basic equation for our circuit:

$$0\,\text{V} - (-15\,\text{V}) = I_B \times 20\,\text{k}\Omega + 0.7\,\text{V} + 2 \times I_E \times 15\,\text{k}\Omega$$

Note the factor of 2 for the current when calculating the voltage drop across R_E. The emitter current of two transistors flows through R_E. If we use $I_E = \beta I_B$ and take a value of $\beta = 300$ we then get:

$$14.3\,\text{V} = (20 \times 10^3 + 2 \times 300 \times 15 \times 10^3) \times I_B = 9.02 \times 10^6 \times I_B$$

and $I_B = 1.58 \times 10^{-6} = 1.58\,\mu\text{A}$ which then readily gives:

$$
\begin{aligned}
V_B &= -20\,\text{k}\Omega \times 1.58\,\mu\text{A} = -0.032\,\text{V} \\
V_E &= -0.032\,\text{V} - 0.7\,\text{V} = -0.732\,\text{V} \\
I_E &= \frac{15\,\text{V} - 0.732\,\text{V}}{2 \times 15\,\text{k}\Omega} = \frac{0.95\,\text{mA}}{2} = 0.475\,\text{mA} \\
V_C &= 15\,\text{V} - 0.475\,\text{mA} \times 10\,\text{k}\Omega = 15\,\text{V} - 4.75\,\text{V} = 10.25\,\text{V}
\end{aligned}
$$

The small signal response is analyzed by noting that a small signal, v_{in}, applied to one input with no signal to the other input is equivalent to a small signal of $+\frac{v_{in}}{2}$ applied to one input and a signal of $-\frac{v_{in}}{2}$ applied to the other input because of the coupling action of the shared emitter resistor. The effect of this signal on one half of the circuit is the same as if that signal were applied to a common emitter amplifier for which the amplification is $-R_C \times \frac{I_E}{25\,\text{mV}}$. Thus the change in the voltage at one collector is the effective input signal times the amplification:

$$\Delta V_C = -R_C \times \frac{I_E}{25\,\text{mV}} \times \frac{v_{in}}{2}$$

and the change in the voltage difference between the two collectors is twice this so that the differential voltage amplification for the amplifier is:

$$A_V = -R_C \times \frac{I_E}{25\,\text{mV}}$$

which is the same as the equation for A_V for the common emitter amplifier.

Current mirror. The current in the base of a transistor is a very sensitive function of the emitter-base voltage and it is not normally possible to control the base current by controlling the base voltage. However, the circuit in Figure 34.5 shows one very important example of how this can be achieved.

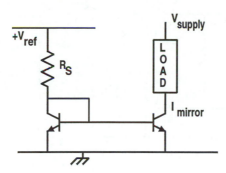

Figure 34.5: Current mirror circuit.

In the left hand transistor the base is connected to the collector to give what is called a diode connected transistor or a transistor which functions as a diode with the voltage drop across the emitter-base junction diode being that appropriate to a current of $\approx \frac{V_{ref}-0.7}{R_S}$. The two transistors are of the same type and in critical applications would be a specially selected matched pair. Since the bases are connected together, the emitter-base voltages are the same and therefore the collector current in the right hand transistor is controlled by the current in R_S in the left hand transistor.

The current in the unknown load driven by the V_{supply} (which may vary) will mirror the current in the left hand transistor. So we have a constant current in the right hand transistor and therefore also in the load. In some circuits the left hand transistor, which is a diode connected transistor, is replaced by a diode or shown on the circuit diagram as a diode but better thermal stability is obtained by using a diode connected transistor.

Tuned amplifier. The gain of a common emitter amplifier is given by $-R_C \times \frac{I_E}{25\,\text{mV}}$ which is essentially independent of frequency. However, coupling capacitors limit the low frequency response and stray capacitances and carrier transit times in the base limit the high frequency response. Between these limits the response of the common emitter amplifier has an essentially flat frequency response curve. In other words, the amplification does not vary with frequency in this central region.

If the R_C is replaced by a resonant circuit, having a maximum impedance at one particular frequency, then the resulting amplifier also has a highly peaked gain versus frequency response curve.

The usual configuration is a parallel LC load such as is shown in Figure 34.6 (a). The Q factor of the LC circuit (see Units 13 and 17) determines the sharpness of the response which is shown in Figure 34.6 (b). The peak

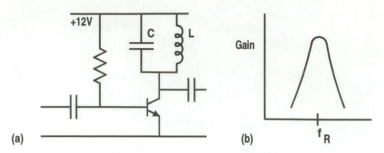

Figure 34.6: Tuned amplifier.

in the gain is at the resonant frequency of:

$$f_r = \frac{1}{2\pi\sqrt{LC}}$$

Calculation of the value of the gain at the maximum is not so straightforward as it depends on the resistance of the inductance and also on the input impedance of the next stage or the load. This type of resonant tuned amplifier allows one particular radio frequency signal, picked up from an antenna, to be selectively amplified and converted, either by amplitude or frequency demodulation, to an audio signal in a radio receiver. The selection of the received station is achieved by varying the resonant frequency of the resonant circuit by using a variable capacitor for tuning. Signals at other frequencies from other transmitters are not amplified and thus do not appear at the final output stage. A number of tuned amplifier stages must be used in a practical radio receiver, if good selectivity is to be obtained.

34.1   Problems

34.1 A signal of $+19\,\text{mV}$ is applied to input 1 and a signal of $+34\,\text{mV}$ is applied to input 2 of the amplifier in Figure 34.4. Calculate the resulting change in the voltage difference between the collectors.

34.2 The input signals used in Problem 34.1 are replaced by $-6\,\text{mV}$ and $+4\,\text{mV}$. Calculate the new collector voltages.

Unit 35   Combining circuit blocks

- A current mirror block can be used as a collector resistor active load in a common emitter amplifier to give a large effective R_C and high gain.

- A current mirror block can be used as the shared emitter resistor in a differential amplifier to give a constant current.

- A common emitter amplifier with a feedback capacitor connected from the collector to the base acts as a low pass filter with a corner frequency proportional to $\frac{1}{C_F}$.

There are only a limited number of basic circuit building blocks which are used in analog electronics and we have analyzed the more important of these in the last three units.

The problem solving power of analog electronics results from combining a small number of well understood circuit blocks to make complex systems. In analyzing a complicated circuit diagram you should always try to split the circuit up into smaller blocks which have well known characteristics. In this unit we will show how the one and two transistor blocks which we have analyzed can be combined to synthesize an operational amplifier and how these blocks then facilitate the fabrication of operational amplifiers as integrated circuits containing about 40 transistors on a single integrated circuit.

We have already analyzed the operation of the common emitter amplifier and found the gain to be given by $-R_C \times \frac{I_E}{25\,\text{mV}}$. This gives a small signal gain of about -200 in a typical common emitter amplifier. If R_C is increased in value, while maintaining I_E constant, the gain is then increased in proportion but the supply voltage must also increase to keep $V_{sup} - I_C \times R_C > 0$. The ideal situation would be an R_C which is large for small signals but which is small for purposes of calculating the DC currents and voltages. This can be achieved by using what is called an active load.

Consider the transistor output characteristic shown in Figure 35.1. The characteristic curves are not horizontal but have a small upward slope.

The curve for $I_B = 10\,\mu A$ increases by $0.2\,\text{mA}$ for a change of $20\,\text{V}$ in V_{CE} to give a dynamic resistance of $100\,\text{k}\Omega$. Therefore a current mirror constructed from transistors having these characteristics is not perfect but

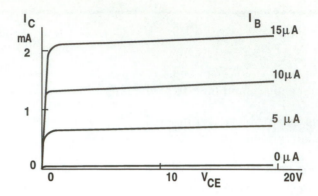

Figure 35.1: Slope of output characteristic gives a high resistance.

does show some small dependence on the V_{CE}. We take advantage of this large dynamic resistance of the current mirror and use it as a collector load resistor in a common emitter amplifier. Effectively we obtain a large R_C for small signal amplification without the need for a larger supply voltage to maintain the I_E in $-R_C \times \frac{I_E}{25\,\text{mV}}$. This is achieved in practice by using a pnp transistor current mirror as a collector resistor in an npn common emitter amplifier so as to obtain a very high gain for small signals. The common emitter amplifier circuit is shown in Figure 35.2 (a).

The common emitter amplifier with the R_C replaced by the current mirror circuit is shown in Figure 35.2 (b).

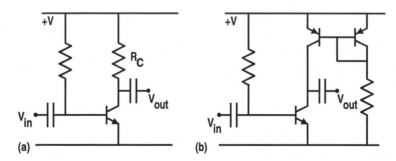

Figure 35.2: Use of a pnp current mirror in place of the collector resistor.

In the differential amplifier discussed in Unit 34, the voltage across the shared emitter resistor, R_E, is nearly constant. This ensures that an increase in current in one transistor of the differential amplifier is balanced by a reduction of the current in the other transistor. This is the circuit which is shown in Figure 35.3 (a). The balancing is improved if the R_E is replaced by a current mirror as shown in Figure 35.3 (b) which gives a more constant current which is to be divided between the two transistors. The total current

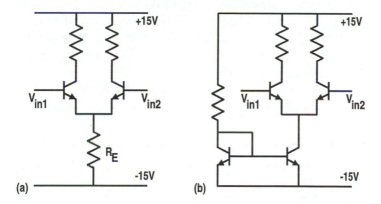

Figure 35.3: Current mirror used as emitter resistor in differential amplifier.

is now also much less dependent on any common mode signal applied to both bases of the transistors together. The circuit is also much less dependent on the value of the negative voltage supply.

If a capacitor is connected from the collector to the base in a common emitter amplifier, as shown in Figure 35.4, the amplified signal at the collector is fed back through the C_F to the base. This amplified and inverted signal

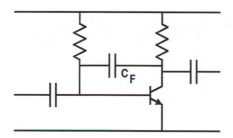

Figure 35.4: Use of collector-base capacitance to reduce high frequency gain.

is added to the input signal and gives a reduced net signal at the base and therefore a reduced output signal. Effectively the gain of the amplifier is reduced. The higher the signal frequency the lower will be the impedance of the capacitance through which the signal is fed back and the smaller will be the amplifier gain. Thus we expect the curve of gain as a function of frequency to show a decrease in gain as the frequency increases.

Even if there is no capacitor connected, there is always some stray capacitance between the collector and the base and so there will always be a decrease in the gain of a common emitter amplifier at high frequencies. Even though the stray capacitance is small its value has to be multiplied by a factor of $(1 + A)$ since it is the amplified signal which appears across the capacitance. This is called the Miller effect and it is a disadvantage if we are trying to make an amplifier which works up to high frequencies. On the other hand, if we want to limit the gain at high frequencies, we increase the effect by connecting a capacitor between the collector and the base to form an effective low pass filter with a corner frequency which depends on C_F and on the input resistance of the common emitter amplifier.

35.1   Problem

35.1 The circuit shown in Figure 35.5 is a schematic version of the circuit which is used in a 741 integrated circuit operational amplifier. Identify each of the circuit block types which are used in the circuit and draw boxes to show which components constitute each circuit block. You should be able to identify five current mirrors, one differential amplifier, one common emitter amplifier, one low pass filter, one emitter follower and one push-pull emitter follower.

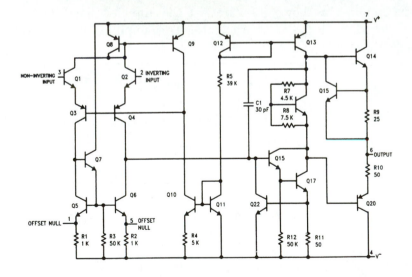

Figure 35.5: Schematic of the internal circuitry of a 741 operational amplifier. Courtesy of National Semiconductors.

Unit 36   Junction field effect transistors

In the semiconductor fabrication process:

- Silicon dioxide is an insulator.

- Silicon dioxide can be formed on silicon substrates.

- Etchant preferentially dissolves silicon dioxide.

- Photoresist protects silicon dioxide from being etched.

- Silicon dioxide prevents dopants from reaching the silicon.

In JFET operation:

- Increasing drain to source voltage gives increased drain current.

- Increasing drain voltage pinches off the conduction channel and reduces the drain current.

- These two processes partially cancel out.

A JFET is specified by three main parameters:

$$
\begin{aligned}
V_{GS(off)} &= & &\text{Gate to source cutoff voltage} \\
I_{DSS} &= & &\text{Drain saturation current for zero gate to source voltage} \\
g_m &= \frac{dI_D}{dV_{GS}} &= &\text{Mutual conductance}
\end{aligned}
$$

Fabrication processes. Before examining how a junction field effect transistor operates, we consider some of the unit operations which are used in the fabrication process and see how these operations are combined in sequences in the fabrication of bipolar transistors and FETs.

A silicon wafer is a 0.5 mm thick, 100 mm, 150 mm or 200 mm diameter disk of silicon sliced, parallel to a crystal plane, from a large, raw crystal or boule of high purity, lightly doped p-type or n-type silicon. This wafer is the substrate on which transistors, FETs and integrated circuits are fabricated.

169

The surface layer of the silicon wafer is usually coated with a thin native oxide layer of SiO_2 about 0.1 nm thick. The thickness of this layer of silicon dioxide can be increased by an oxidizing process in which the wafer is heated to a temperature of the order of 1000 K in an atmosphere of steam and oxygen which results in the oxidation of the surface layer of the silicon wafer to silicon dioxide. Other methods of oxidation include electrochemical anodization and plasma reaction.

Instead of oxidizing the silicon of the wafer, a silicon dioxide layer can also be formed on the wafer by a chemical vapour deposition process in which silane (SiH_4) gas is reacted with oxygen gas to leave a deposit of silicon dioxide on the wafer surface. This method is used for forming thicker layers of silicon dioxide.

The silicon dioxide, which is essentially a layer of glass, can be dissolved by etching the wafer in hydrogen fluoride (HF). The HF does not dissolve the silicon.

A photoresist is an organic material that can be coated or spun on to the wafer surface to give a layer about a micron thick. When a negative type photoresist is exposed to ultraviolet light, it polymerizes and hardens. In the developing process, unexposed, unpolymerized photoresist is dissolved and removed. The exposure of the photoresist coated wafer is carried out using a photomask which is a patterned layer of metal on a glass substrate. The UV light passes through the clear areas and exposes the photoresist. The UV light does not pass through the metallized regions of the photomask. After the development process, a hardened resist pattern coats the wafer with a pattern which is the negative of the metal pattern on the mask.

In the next stage of the process the oxide layer in the regions not protected by hardened photoresist is etched away either by immersing the wafer in HF or by use of a plasma etcher. This forms windows in the oxide in the same pattern as the original metal pattern on the mask. When the process is completed the remaining photoresist can then be removed by a plasma ashing process or by a wet chemical process.

The wafer is then placed in a furnace in an atmosphere of either p-type (boron) or n-type (phosphorous) dopant containing gas. The dopant enters the silicon wafer through the windows in the oxide and diffuses down into the wafer. The wafer is protected from the dopant gas in the other regions by the oxide. The sequence of operations is illustrated in Figure 36.1.

If the remaining parts of the oxide layer are etched away, a new oxide layer can be grown on the wafer and the process can be repeated with a different mask pattern and multiple structured layers can be formed in the top layers of the silicon wafer. Metal layers and layers of polysilicon can be evaporated or deposited onto the wafer and etched in patterns in exactly the same way

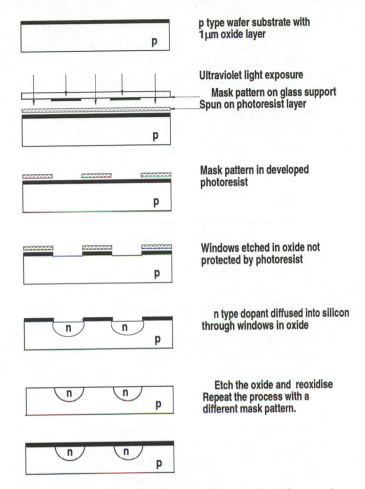

Figure 36.1: Photolithography using a negative photoresist.

as the silicon but using other etchants besides HF. Connections can be made to specific doped regions of the pattern and the connecting tracks can cross each other if silicon dioxide is used as an insulating layer between the tracks. Capacitors can be formed on the the wafer by placing an oxide layer between two metallic layers. If the dopant concentrations are tightly controlled and the n-type or p-type tracks are long then it is also possible to form resistors on the wafer.

Since the mask can contain multiple copies of the same small mask pattern, it is possible to form many thousands of individual transistors or FETs or many hundreds of integrated circuits on a single wafer. At the end of the process, the wafer is scribed and diced and the individual devices can then be separated, tested, packaged and sold. Since the tracks in the masked patterns can be as small as microns it is important that dust particles, which might give unwanted additional masking, be kept away from the process. This is the reason why the process is carried out in superclean rooms by operators wearing dust free suits. Even with all these precautions, the typical yield, which is of the order of 99% for single transistor manufacture, can drop to 50% for complex integrated circuits such as microprocessor chips or memory chips containing up to a million transistors in a single integrated circuit. Failure of a single transistor of the thousands of transistors in an IC causes the whole IC to be failed at the inspection stage.

Bipolar transistors. We have already examined the operation of npn and pnp bipolar transistors. The fabrication of these transistors can, in principle, be carried out using two masks for the diffusion stages with the masks consisting of a simple metal circle pattern. The diameter of the metal is smaller on the second mask. The result of masking, etching and doping an n-type substrate with p-type and then n-type gives the planar structure shown in Figure 36.2 which is an npn transistor. Note that the emitter region started out as n-type substrate, it was then doped p-type during the base region formation and was turned back to n-type by further doping during the emitter region formation.

The connections can be made using thin gold wire welded to metal pads on the base and emitter and to the lead-out wires in the packaging can. The collector, which is the substrate, is often bonded to the metal of the can which then forms the third connection. This is why many transistors have the can connected to the collector. When the collector is bonded to the can, the heat sinking of the silicon transistor is improved since there is good thermal contact between the silicon and the metal can. If there is significant power dissipation in the transistor, the can is often mounted on a heat sink with radiating fins to keep it cool. It can also be seen from Figures 36.2 and 36.3 that the model of an npn bar shaped transistor, with connections at the side, which we used at the beginning of Unit 30, really corresponds to a narrow vertical section of a much wider and flatter structure.

In the microphotograph shown in Figure 36.3, the top of the metal can of a BC109 transistor has been cut off and the silicon chip, the doped areas and the connecting wires are shown. The chip has dimensions about $0.5\,\text{mm} \times 0.5\,\text{mm}$. The small diameter of the wires should be noted. If too large a current is passed through the transistor the connecting wires of small signal transistors tend to heat and fuse giving an open circuit on the lead. This is why simple tests for diode action can usually verify the functioning of signal transistors. The leads of power transistors are usually heavier and can last for long enough for the silicon to overheat and change the performance

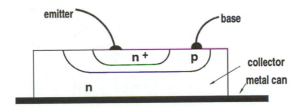

Figure 36.2: Cross section of an npn transistor.

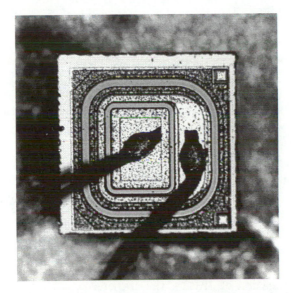

Figure 36.3: Microphotograph of a BC109 transistor.

of the transistors. As a result, more complex checking procedures may be necessary for suspect power transistors in a circuit since they may degrade rather than fail catastrophically.

Junction field effect transistors. The unit processes involved in the manufacture of JFETs are the same as those involved in the manufacture of bipolar transistors. The processes are compatible and it is possible to manufacture JFETs and bipolar transistors on the same wafer in integrated circuits. However, a different arrangement of masks is used to give a different geometry for JFETs.

A top view of an n channel JFET is shown in Figure 36.4. The conduction path between the drain and source is along the n channel in the n-type epilayer between the gate and the p-type substrate as shown in the section view of the JFET. If the annular structure shown in Figure 36.4 is

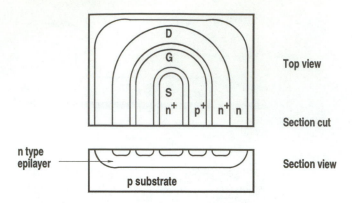

Figure 36.4: Structure of an n channel JFET.

used, it is then possible to fabricate JFETs and not have the gate connected to the substrate. This is an advantage because it allows a metal can encapsulation to be used without the gate being connected to the can. Use of a plastic encapsulation allows this problem to be avoided. The annular structure also allows a number of JFETs to be fabricated on one chip without all of the gates being shorted together so that JFETs can then be combined in integrated circuits on one chip.

The essential difference between the bipolar transistor and the JFET is that the current flows perpendicularly through the thin base region in the bipolar transistor but along the channel in the case of the JFET.

In the fabrication of JFETs, the starting point is a p-type substrate on which an n-type circular region is formed. The central heavily doped n-type source region and outer drain annulus are formed next. Then the heavily doped p-type gate annulus is formed between the drain and source. The n-type channel extends under this gate region and above the p-type substrate. In analyzing the operation of this n channel JFET, we consider a radial section through the device, shown schematically in Figure 36.5.

The depletion layer associated with the heavily doped p-type gate region extends into the channel. As the reverse bias applied to the gate region increases, the channel thickness is constricted. Therefore the application of reverse bias voltage to the gate controls the current flowing in the channel between the drain and source. The JFET is then a voltage controlled device whereas the bipolar transistor is a current controlled device.

In many JFETs, the p-type gate is connected to the p-type substrate and the channel is constricted between the gate and the substrate with a depletion layer above and below.

If this were the full story, the structure would simply behave as a resistor

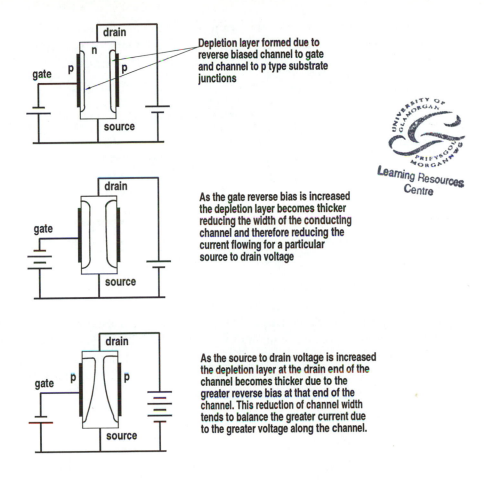

Figure 36.5: Mechanism of operation of a JFET.

between the drain and source having a resistance value determined by the voltage applied to the gate.

However, another mechanism also operates. For voltages greater than what is called the pinch-off voltage, V_P, applied between the drain and the source, the increased reverse bias between the upper end of the gate and the drain gives a thicker depletion layer at the top of the channel. This is shown in Figure 36.5. Thus the increased drain voltage, besides driving more current through the channel, also causes a reduction of the channel width which tends to counteract the increased current which would otherwise flow between the drain and the source. Therefore if the gate voltage is kept constant and the drain to source voltage is increased, it is found that the drain current initially increases in proportion to the drain to source voltage but that it then levels off so that the drain current is nearly independent of the drain to source voltage. The characteristic which results from this pinch-off effect is shown in Figure 36.6.

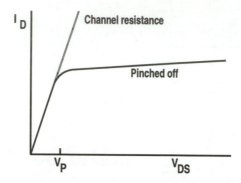

Figure 36.6: Effect of the pinch-off mechanism.

In order to characterize a JFET, a circuit such as that shown in Figure 36.7 is used. This circuit shows an n channel JFET such as a 2N3819 device. The diode arrow on the gate of the JFET symbol indicates the direction of the pn junction diode between gate and channel. The diode arrow for a p channel JFET points in the opposite direction.

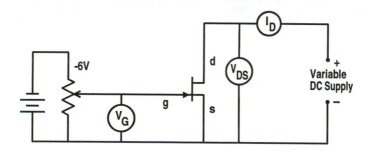

Figure 36.7: Circuit for measuring JFET characteristics.

This circuit allows us to apply a negative voltage between $0\,V$ and $-6\,V$ to the gate of the JFET, measured with reference to the source. Since the gate to channel is a reverse biased diode there will be no significant current in the gate lead. The circuit also allows us to measure simultaneously the drain to source voltage and the drain current.

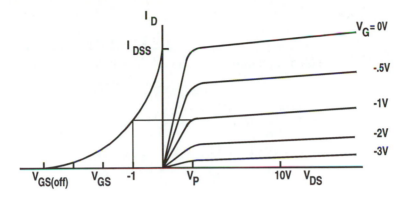

Figure 36.8: JFET characteristics.

Figure 36.8 shows the characteristic curves which result from these measurements. In the right hand set of curves, the drain current is plotted as a function of the drain to source voltage for various constant values of the gate to source voltage. The general shape is a rising region followed by a saturated region in which the current is pinched off. For a gate to source voltage of $0\,\text{V}$, the drain current is a maximum and this value of the drain current is called I_{DSS}, the drain saturation current for a gate voltage of $0\,\text{V}$. This saturation occurs for a drain to source voltage, $V_{DS} = V_P$, where V_P is the pinch-off voltage.

In the left hand curve, we plot the drain saturation current for the particular value of the gate to source voltage as a function of the gate to source voltage. The gate to source voltage at which the drain current drops to $0\,\text{mA}$ is called the gate-source cutoff voltage, $V_{GS(off)}$.

From a theoretical analysis of the physics of JFETs, which we will not pursue in this text, it is found that the I_D at a particular value of V_{GS} is given by:

$$I_D = I_{DSS}\left(1 - \frac{V_{GS}}{V_{GS(off)}}\right)^2$$

which is the theoretical equation for the curve of I_D versus V_{GS} in the left hand characteristic curve in Figure 36.8. The slope of this curve is the mutual conductance, g_m, for the JFET. It can be seen that this is not a constant but depends on the operating point. The data sheets usually quote a value of g_m obtained at some typical JFET operating value. The quoted value for g_m for the 2N3819 JFET is $g_m = 2000\,\mu\text{S}$ (microsiemens) at $V_{DS} = 5\,\text{V}$. The siemen is the unit of conductance and has units of $\frac{\text{Amps}}{\text{Volts}}$.

36.1   Example

36.1 If $I_{DSS} = 7.5\,\text{mA}$ and $V_{GS(off)} = -3.7\,\text{V}$ for the JFET characteristics plotted in Figure 36.8, calculate the value of I_D when $V_{GS} = -1\,\text{V}$.

$$
\begin{aligned}
I_D &= 7.5 \times \left(1 - \frac{-1}{-3.7}\right)^2 \text{mA} \\
&= 7.5 \times (1 - 0.27)^2 \text{mA} \\
&= 3.99\,\text{mA}
\end{aligned}
$$

36.2   Problems

36.1 The following values were quoted in the data sheet for an n channel JFET:

$$ g_m = 2700\,\mu\text{S}, \quad V_{GS(off)} = -4.5\,\text{V} \quad \text{and} \quad I_{DSS} = 7.0\,\text{mA} $$

Reconstruct the characteristic curves for this JFET from this set of values. (See Figure 36.8.)

36.2 If $I_{DSS} = 9.3\,\text{mA}$ and $V_{GS(off)} = -3.7\,\text{V}$ for a JFET having the characteristic shown in Figure 36.8 calculate the value of g_m at $V_{DS} = 10\,\text{V}$.

36.3 Draw up a set of masks suitable for use in fabricating the BC109 transistor shown in Figure 36.3. Use negative photoresist.

36.4 What changes would be necessary if positive photoresist was used in Problem 36.3?

36.5 Modify the circuit of Figure 36.7 so that the characteristics of a p-channel JFET can be measured.

Unit 37   JFET amplifiers

- To autobias a JFET amplifier, select: $R_S = \left| \frac{V_{GS(off)}}{I_{DSS}} \right|$

- This gives a gate to source voltage of: $V_{GS} \approx 0.4 \times V_{GS(off)}$

- The drain current is then: $I_D \approx 0.4 \times I_{DSS}$

- The common source small signal gain is: $A_V = -g_m \times R_D$

- Typical small signal gains are of the order of 10.

- Typical input impedances are of the order of $1\,\text{M}\Omega$.

If you consult the components catalog of a typical electronic component supplier, you will find that the only readily available parameters for a junction field effect transistor, such as the 2N3819, are I_{DSS}, $V_{GS(off)}$ and g_m together with the maximum voltage and current ratings. A full set of characteristic curves may be available from the component manufacturer but obtaining them can take time. We have to be able to use a JFET even when a full set of characteristic curves is not readily available or without having to reconstruct them from the available data. (See Problem 36.1.)

Another difficulty is that the catalog values for I_{DSS}, $V_{GS(off)}$ and g_m are average values for that JFET type number. A particular JFET, taken out of the storage bin at random, may have parameter values significantly different from the catalog values and still be within the acceptable product specification spread. It is therefore necessary to have a quick method of measuring the parameters for a particular JFET before using it.

The circuit shown in Figure 37.1 (a) shows how the I_{DSS} can be measured directly. Note that the gate is connected to the source to give $V_{GS} = 0\,\text{V}$. The circuit in Figure 37.1 (b) gives a very small value of I_D since the current through the $20\,\text{k}\Omega$ gives a V_{GS} so close to the cutoff voltage that it can

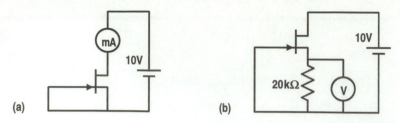

Figure 37.1: Circuits to measure I_{DSS} and $V_{GS(off)}$ for a JFET.

be taken to be equal to the gate-source cutoff voltage. These two simple measurements give I_{DSS} and $V_{GS(off)}$ for the particular JFET in use.

The equation for JFET drain current at any gate to source voltage is:

$$I_D = I_{DSS} \times \left(1 - \frac{V_{GS}}{V_{GS(off)}}\right)^2$$

Differentiate this to get:

$$g_m = \frac{dI_D}{dV_{GS}} = -2\frac{I_{DSS}}{V_{GS(off)}}\left(1 - \frac{V_{GS}}{V_{GS(off)}}\right)$$

Note that g_m is positive since both V_{GS} and $V_{GS(off)}$ are negative quantities.

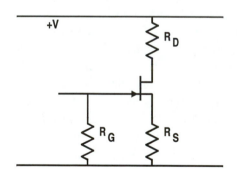

Figure 37.2: JFET amplifier autobias circuit.

Before using a JFET as an amplifier it is necessary to bias the JFET correctly. Consider the simple autobias circuit, shown in Figure 37.2, in which the source current through the JFET also flows through R_S to give a reverse bias of $I_D \times R_S$ between the gate and source. The gate voltage is 0 V because no significant current flows through R_G and through the reverse biased gate to channel junction.

There is a wide range of possible values for R_S which will give a reasonable bias configuration. The most straightforward procedure for selecting the value of R_S is to use:

$$R_S = \frac{V_{GS(off)}}{I_{DSS}}$$

This causes the bias point to be at the intersection of the line for R_S and the curve for V_{GS} shown in Figure 37.3.

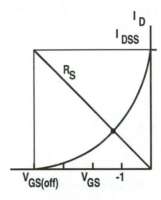

Figure 37.3: Method of selecting R_S.

We then have $V_{GS} = I_D \times R_S = I_D \times \frac{V_{GS(off)}}{I_{DSS}}$ which gives:

$$\frac{V_{GS}}{V_{GS(off)}} = \frac{I_D}{I_{DSS}}$$

But we also have the equation for I_D for the JFET:

$$\frac{I_D}{I_{DSS}} = \left(1 - \frac{V_{GS}}{V_{GS(off)}}\right)^2$$

Substitute $\frac{V_{GS}}{V_{GS(off)}}$ for $\frac{I_D}{I_{DSS}}$ in this equation to get an equation of the form $x = (1-x)^2$ which has solutions $x \approx 2.6$ and $x \approx 0.4$.

Take the only physically possible solution, $x \approx 0.4$, to get:

$$\frac{V_{GS}}{V_{GS(off)}} = \frac{I_D}{I_{DSS}} = 0.4$$

This method of biasing the JFET then gives a drain current of:

$$I_D = 0.4 \times I_{DSS}$$

By following this procedure, a reasonable bias point is obtained for the simple autobias circuit. However, you should remember that this is just one bias point out of a range of possible bias points so you will find other values used in circuits.

The next step in the design of a JFET amplifier is to select the value for the drain resistance, R_D.

If a small signal is coupled to the gate through a capacitance, the operating point is momentarily displaced from the DC bias point. The source voltage can be maintained constant by the use of a capacitor connected across the source resistance. This configuration is shown in Figure 37.5.

The mutual conductance, g_m, relates the change in drain current to a change in the gate voltage:

$$g_m = \frac{dI_D}{dV_{GS}} = \frac{i_d}{v_g}$$

The drain resistor relates the change in drain current to the change in the output voltage:

$$v_d = -R_D \times i_d$$

Combine these two relationships to get the small signal voltage gain:

$$A_V = \frac{v_{out}}{v_{in}} = \frac{v_d}{v_g} = \frac{-R_D \times i_d}{\frac{i_d}{g_m}} = -g_m R_D$$

If the required gain, A_V, is specified, the value for R_D is calculated from:

$$R_D = \frac{-A_V}{g_m}$$

The load line can now be drawn as R_D on the characteristic curve shown in Figure 37.4. It is necessary that this load line intersects the horizontal

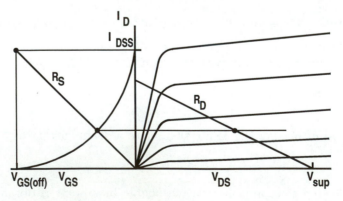

Figure 37.4: Identification of required load line.

line corresponding to $I_D = 0.4 \times I_{DSS}$ so we then obtain a minimum required value for the power supply voltage. Typically we require that the load line intersects the I_D axis at a point well above $0.4 \times I_{DSS}$. It can be seen from Figure 37.4 that if the same value of R_D were to be used with a smaller supply voltage, the load line and the horizontal line for I_D would not intersect within the range of the characteristic curves.

If the load line of slope R_D is drawn through the point $2 \times 0.4 I_{DSS} = 0.8 I_{DSS}$ on the I_D axis then there is enough available voltage across R_D to allow the drain voltage to swing equal amounts in the $+$ and $-$ directions. This criterion then allows an optimum supply voltage to be selected as:

$$V_{sup} = 0.8 \times I_{DSS} \times R_D$$

Since g_m is of the order of $2000\,\mu$S and R_D is typically of the order of $5\,\text{k}\Omega$, small signal gains of the order of $g_m \times R_D = 10$ are expected for JFET amplifiers. This is much lower than the gains of approximately 200 which are obtained with bipolar transistor amplifiers. However, the JFET amplifier has the advantage that the input impedance is much higher and is set by the resistance connected to the gate.

37.1 Example

37.1 Calculate the component values and supply voltage for the circuit in Figure 37.5, so as to obtain a small signal voltage amplification, $A_V = -6$ and an input impedance of $500\,\text{k}\Omega$.

The JFET used in the circuit has the following parameter values: $g_m = 2500\,\mu$S, $V_{GS(off)} = 2.5\,$V and $I_{DSS} = 7.5\,$mA.

(In this example, the calculated component values have been inserted in the diagram.)

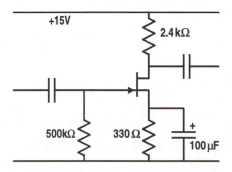

Figure 37.5: Circuit for Example 37.1.

Since R_{in} is to be $500\,\text{k}\Omega$ and the input gate resistance of the JFET is $\gg 1\,\text{M}\Omega$, we pick $R_G = 500\,\text{k}\Omega$ to give the required input impedance for the amplifier.

We obtain a reasonable value for R_S by setting:

$$R_S = \frac{V_{GS(off)}}{I_{DSS}} = \frac{2.5\,\text{V}}{7.5\,\text{mA}} = 330\,\Omega$$

If we draw a simplified set of generic JFET characteristics as shown in Figure 37.6, we can see the approximate position of the drain current.

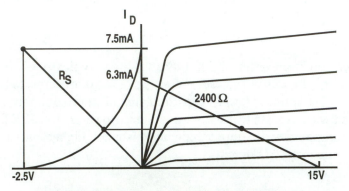

Figure 37.6: Load line for Example 37.1.

We need a small signal gain of -6. Therefore:

$$A_V = -6 = -g_m \times R_D = -2500 \times 10^{-6} \times R_D$$

This gives $\quad R_D = \frac{6}{2500 \times 10^{-6}} = 2400\,\Omega = 2.4\,\text{k}\Omega$

This load line is sketched in Figure 37.6 (note that the voltage drop across R_S has been neglected). The line for the drain current of $0.4 \times 7.5\,\text{mA} = 3\,\text{mA}$ from the R_S selection calculation must intersect the load line for R_D in the central region of the JFET characteristics. It can be seen that a supply voltage of $15\,\text{V}$ gives a reasonable intersection point.

37.2   Problems

(The catalog values for the JFETs used in these problems are: $g_m = 2000\,\mu\text{S}$, $V_{GS(off)} = -4\,\text{V}$ and $I_{DSS} = 9\,\text{mA}$).

37.1 If the power supply voltage in Example 37.1 is reduced from $+15\,\text{V}$ to $+6\,\text{V}$, will the amplifier still be biased correctly?

37.2 A JFET is selected at random from the storage bin. How would the I_{DSS} and the $V_{GS(off)}$ for the JFET be measured?

37.3 If the JFET in the circuit in Figure 37.5 is replaced and the new value of the source voltage is found to be $V_S = 1.4\,\mathrm{V}$, calculate the new drain current and the new drain voltage.

37.4 Calculate the component values and supply voltage for the circuit in Figure 37.7 which give a small signal voltage amplification, $A_V = -7$, and an input impedance $R_{in} = 1\,\mathrm{M\Omega}$.

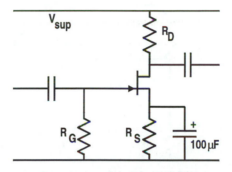

Figure 37.7: Problem 37.4.

37.5 Sketch the voltage waveforms which you would observe with an oscilloscope connected to the input, the gate, the source, the drain and the output in the circuit of Problem 37.4 when a $1\,\mathrm{kHz}$ sinusoidal signal of amplitude $70\,\mathrm{mV}$ is applied to the amplifier input.

37.6 Calculate suitable values for the resistors for the source follower circuit shown in Figure 37.8. What would be a reasonable value for the supply voltage?

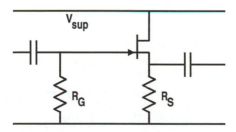

Figure 37.8: Problem 37.7.

37.7 What is the time constant for the filter formed by the R_S and the 100 μF in the circuit of Problem 37.4?
What is the corner frequency corresponding to this time constant?
Will the design voltage gain of -7 be obtained for frequencies below this corner frequency?

37.8 Calculate suitable values for the source resistor, R_S, in the circuit shown in Figure 37.9. (Hint: Calculate the I_D as in Problem 37.4 and then increase the value of R_S to allow for the increased voltage corresponding to the voltage at the mid point of the potential divider.)

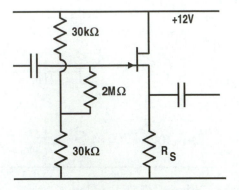

Figure 37.9: Problem 37.8.

37.9 Calculate the drain current for the circuit in Figure 37.10. Calculate the source, drain and gate voltages.

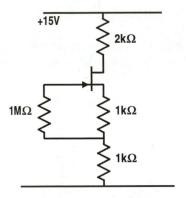

Figure 37.10: Problem 37.9.

Unit 38 MOSFET and CMOS

- The depletion type MOSFET conducts for a gate to source voltage of zero.
  It is a **normally on** device.
  Drain current drops to zero at $V_{GS(off)}$.

- The enhancement type MOSFET does not conduct for a gate to source voltage of zero.
  It is a **normally off** device.
  The device starts to conduct at threshold voltage $V_{GS(th)}$.

- CMOS devices use complementary p and n channel enhancement type MOSFETs in one integrated circuit of two or more MOSFETs.

MOSFET fabrication. The starting point for MOSFET (Metal Oxide Silicon Field Effect Transistor) fabrication is a lightly doped silicon wafer with a surface layer of silicon dioxide. The photoresist, etch and diffusion processes described in Unit 36 are used to form two heavily doped, isolated regions or wells on the wafer. In the n channel depletion type MOSFET these wells are n-type. An n-type channel is then formed which links the source and drain n^+ wells. (n^+ denotes heavily doped n-type.) In the n channel enhancement type MOSFET this n-type doped channel is not fabricated.

A very thin layer of silicon dioxide is formed over the channel region and a metallized gate region is formed on the silicon dioxide over the channel. A thick silicon dioxide layer is deposited over the device and windows are etched through this silicon dioxide to permit metal contacts to be made to the n^+ source and drain wells and to the gate metallization. It is a fundamental feature of MOSFETs that there is no conducting path between the gate lead and the channel. The gate is isolated by the thin silicon dioxide layer (effectively a glass insulator) which separates it from the channel. This is why you will sometimes see the MOSFET referred to as an IGFET (Insulated Gate Field Effect Transistor).

Since there is an insulating oxide layer between the gate and the channel in the MOSFET, the source, drain and channel can be fabricated as a linear or rectangular structure with the gate overlaying the channel but insulated

187

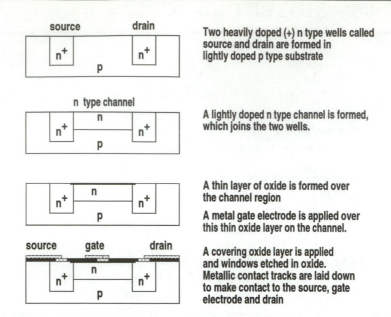

Figure 38.1: Stages in the fabrication of an n channel MOSFET.

from it. The annular structure used in the JFET is no longer necessary in order to prevent a short between the gate and substrate. This means that the fabrication masks for MOSFETs can be rectangular strips, which allows a much simpler fabrication process to be used and which also gives a higher device packing density. This is one of the reasons why the MOSFET technology is so widely used for high density computer chip and memory chip manufacture.

This insulating layer gives the MOSFET its very high input resistance which is in excess of $10^9 \, \Omega$. However, this thin silicon dioxide layer is also the weak point in MOSFETs. If the MOSFET is handled without taking precautions to avoid static, the high voltages associated with static (a few thousand volts) can cause a spark discharge to punch through the thin silicon dioxide in the gate region and destroy or shorten the life of the MOSFET.

The depletion mode MOSFET. Depletion mode MOSFET characteristic curves are shown in Figure 38.2 and are similar to those for the JFET except that, because of the insulated gate, it is now permissible to operate the device with either a negative or a positive bias voltage on the gate. The gate voltage at which the drain current is cut off is called the gate-source cutoff voltage, $V_{GS(off)}$ or V_{off}.

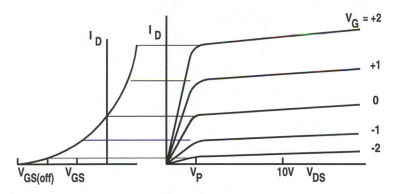

Figure 38.2: n channel depletion mode MOSFET characteristic curves.

If the voltage applied to the gate is negative, the MOSFET operates in depletion mode and the gate electrode voltage repels the channel electrons from the region under the gate electrode, reducing the width of the conducting channel and reducing the drain current. The pinch-off mechanism which we discussed in the case of the JFET also operates whereby increased drain voltage pinches the channel to give a nearly constant drain current.

If the gate voltage is positive, the MOSFET operates in enhancement mode and electrons are attracted into the channel by the positive gate voltage to give increased drain current. There is a second mechanism which can be considered to operate. Holes in the substrate are repelled from under the gate by the positive voltage applied to the gate and this leads to an increase in the number of electrons in the channel region because of the semiconductor equation, $n \times p = n_i^2$ (see Unit 22). If p in $n \times p = n_i^2$ decreases then n must increase.

The main advantage of being able to put voltages of either polarity on the gate is that it is possible to use circuits such as that shown in Figure 38.3 (a) where no reverse bias is applied to the gate but the gate is allowed to vary on either side of zero volts. Figure 38.3 (b) shows the variation of the gate voltage and the resulting variation of the drain current. The fact that a source resistor, R_S, is not needed is not a significant saving in a single circuit but if MOSFETs are used in integrated circuits containing many hundreds of such amplifiers in a single circuit block then the saving can be significant.

Note the circuit symbol used for the depletion mode, n channel MOSFET used in the circuit. The channel is represented by a continuous bar, the gate is isolated from the bar and the fact that the channel is n-type is indicated by the direction of the diode arrow from the substrate to the channel. A p-type channel MOSFET has the diode arrow pointing in the other direction.

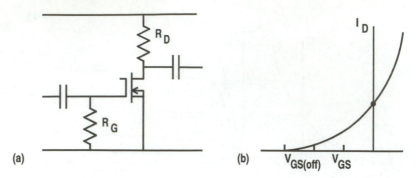

Figure 38.3: Basic MOSFET amplifier circuit and input characteristic.

The enhancement mode MOSFET. The n channel enhancement mode MOSFET is fabricated without a doped n-type channel connecting the n$^+$ source and drain wells. The application of a large enough positive voltage, called a threshold voltage, $V_{GS(th)}$, to the gate causes an n-type inversion layer to be formed in the p-type region between the source and the drain due to the electrostatic repulsion of the positive gate voltage acting on the p-type carriers. The semiconductor equation, $n \times p = n_i^2$, then gives an increased electron concentration which results in the formation of an n-type channel connecting the source to the drain. The positive gate voltage enhances the conduction in the channel. The characteristic curves for the device are shown in Figure 38.4.

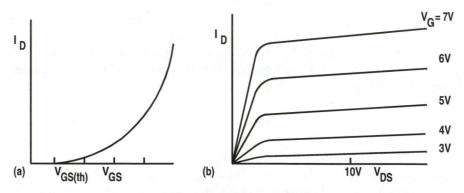

Figure 38.4: n channel enhancement MOSFET characteristics.

It is a fundamental feature of the enhancement type MOSFET that a voltage in excess of the threshold voltage must be applied to the gate before there is any conduction through the device.

Once conduction occurs between the source and drain, the current in the drain in the saturation region is given by the equation:

$$I_D = k(V_{GS} - V_{GS(th)})^2$$

where k is a constant for the particular MOSFET in use.

It is not possible to use the concept of I_{DSS} for an enhancement type MOSFET since there is no current for zero gate to source bias. Another difficulty is that many of the component suppliers' catalogs only quote maximum ratings for the components. It is therefore necessary to have a quick way of determining the MOSFET parameters. The circuit in Figure 38.5 (a) allows the threshold voltage, $V_{GS(th)}$, to be determined since $I_D \approx 0$ and then $V_{GS} = V_{GS(off)}$. In the circuit in Figure 38.5 (b), the drain current is determined from the voltage drop across the drain resistor. This gives an I_D and V_{GS} which can be inserted into the equation for I_D, above, and which will then allow k to be calculated.

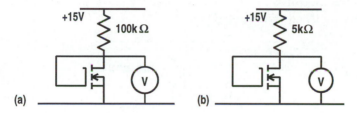

Figure 38.5: Determination of enhancement MOSFET parameters.

Figure 38.6 shows two possible bias configurations for the enhancement type MOSFET. In both circuits, the supply voltage must be greater than the threshold voltage. The circuit in Figure 38.6 (a) has the disadvantage that the gate bias is equal to the supply voltage and this imposes some restrictions on the supply voltage. You should note that the circuit symbol for the enhancement type MOSFET is characterized by an interrupted bar for the channel. The type of channel and the gate are identified by the direction of the diode arrow from the substrate in the same way as in the depletion type MOSFET.

The main advantage of MOSFET devices for analog electronics is the very high input resistance which makes them ideal for input stages of instrumentation devices such as pH meters, multimeters, oscilloscopes, radio receivers etc., which require a high input impedance first stage. Variations of the MOSFET such as the VMOS, the HEXFET, the SIPMOS, in which the channel length is made as short as possible, are used for high current and high frequency operation in applications such as radio transmitters in mobile

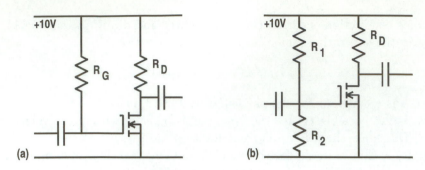

Figure 38.6: Two types of MOSFET bias circuit.

phones, switched mode power supplies and other mobile, portable or battery powered devices.

CMOS. Complementary Metal Oxide Silicon FETS are combinations of p and n channel MOSFETs fabricated side by side on the same wafer. In general the MOSFETs are used as active load type devices where load resistors are replaced by MOSFETs.

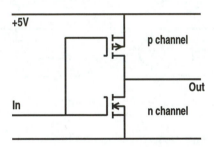

Figure 38.7: CMOS inverter.

A typical circuit configuration is shown in Figure 38.7. A p channel MOSFET is used at the top and an n channel MOSFET is used at the bottom. The two drains are connected together and connected to the output.

Consider the two extreme cases.

If the input voltage to the circuit is at 0 V, then the lower n channel MOSFET gate is below the threshold voltage and the MOSFET is off or nonconducting. However, the gate to source voltage for the upper MOSFET is above the threshold voltage and the upper MOSFET is on and conducting so that the output is connected to the supply voltage.

If the input voltage is at the supply voltage, then the lower MOSFET gate is above the threshold voltage and the lower MOSFET is conducting to give 0 V at the output. The upper MOSFET gate to source voltage is 0 V and therefore the upper MOSFET gate voltage is below threshold and is off and nonconducting.

Effectively, we have an inverter where $0\,\text{V}$ in gives V_{sup} out and V_{sup} in gives $0\,\text{V}$ out. This is a circuit in which one of the MOSFETs is always off so that there is no quiescent current which is what makes this type of circuit suitable for use in battery powered devices.

38.1 Example

38.1 The circuit in Figure 38.8 (a) shows an analog switch which uses an enhancement type MOSFET. The input signal can have any value in the range $-5\,\text{V}$ to $+3\,\text{V}$. Calculate the output voltage for a load resistor of $10\,\text{k}\Omega$ when the control input is at $-5\,\text{V}$ and when the control input is at $+5\,\text{V}$. The threshold voltage for the MOSFET is $+2\,\text{V}$.

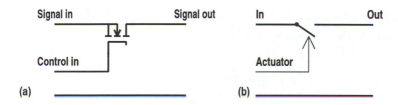

Figure 38.8: MOSFET analog switch.

In this type of analog switch, the control is isolated from the signal which is being switched. The effective switch is shown in Figure 38.8 (b) and is connected between the input and output.

When the control input is at $-5\,\text{V}$, the gate to source voltage is in the range from $0\,\text{V}$ to $-8\,\text{V}$ because the source can be at any voltage within the limits $-5\,\text{V}$ to $+3\,\text{V}$. There is therefore no channel formed between the source and the drain and effectively the switch contacts between the input and output are open and the switch is OFF.

When the control input is at $+5\,\text{V}$, the gate to source voltage is in the range from $+10\,\text{V}$ to $+2\,\text{V}$ for input voltages between $-5\,\text{V}$ and $+3\,\text{V}$. The channel has therefore been formed in the enhancement MOSFET and therefore the resistance between the source and drain is low, typically of the order of $1\,\text{k}\Omega$. The input is then connected to the output and effectively the switch is ON. The output voltage will then be close to the input voltage but there will be some reduction in the voltage due to the potential divider action of the source to drain resistance and the $10\,\text{k}\Omega$ load resistor. In this case the output voltage will be $\frac{10}{11}V_{in}$.

This type of analog switch is available as an integrated circuit (CMOS 4066B type device) which has the feature that the ON resistance is of the order of $50\,\Omega$ with the result that the voltage loss in the device is smaller than we have in this example.

38.2   Problems

38.1 An enhancement type MOSFET has a $V_{GS(th)} = +3\,V$ and also $I_D = 3\,mA$ when $V_{GS} = +5\,V$. Sketch the characteristic curves for the device.

38.2 The drain current of an enhancement type MOSFET is given by:

$$I_D = k(V_{GS} - V_{GS(th)})^2$$

Obtain an expression for g_m for the device.

38.3 The test circuits in Figure 38.5 were used to obtain a $V_{GS(th)} = 2.7\,V$ and an $I_D = 2.3\,mA$ for a $V_{GS} = 4.2\,V$. Calculate the value of k in the equation for I_D.

38.4 Calculate the drain current and voltage for the circuit in Figure 38.9. Use the MOSFET from Problem 38.3.

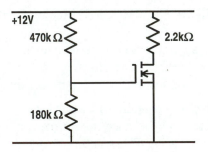

Figure 38.9: Problem 38.4.

38.5 An IRFD110 n channel enhancement type MOSFET is used in an amplifier circuit such as that in Figure 38.6 (b). The amplifier is to have a gain of 40. Calculate suitable values for the resistors in the circuit and sketch the circuit.

The IRFD110 device has $V_{GS(th)} = 3\,V$, $g_m = 1.2\,S$ and $I_D = 2.2\,A$ for $V_{GS} = 6\,V$ and $V_{DS} = 10\,V$.

Unit 39   Operational amplifiers

- The following two rules are used to analyze the operation of op-amps in linear circuits:

  - **Rule 1.** When an op-amp is used in the linear region, the voltage difference between the inverting and noninverting inputs is approximately zero.

  - **Rule 2.** No current flows into the input terminals of the op-amp.

- The gain of an inverting amplifier is given by:

$$A_V = -\frac{R_f}{R_{in}}$$

The term operational amplifier or op-amp is used to describe a directly coupled amplifier fabricated as an integrated circuit on a single silicon chip. There are many types of op-amp manufactured but the most common is the 741 op-amp and this is the one which we will use. Most other op-amp types are improvements on the 741 in terms of stability, frequency response and input impedance at an increased cost.

Figure 39.1: Magnified X-ray views of a 741 op-amp.

We have already met the 741 op-amp internal circuit in Problem 35.1 where you examined the internal circuit blocks of the op-amp. Figure 39.1

shows x, y and z direction X-rays of a 741 op-amp in an 8 pin dual in line (dil) package. The fine wires connecting the silicon chip mounted on the lead frame to the pins of the dil package can just be seen in the left hand X-ray. The silicon chip is mounted on the central square of the lead frame but does not show up on the X-ray as the silicon is transparent to X-rays due to the low atomic number of silicon.

It is not necessary to have a detailed knowledge of the internal circuitry of the 741 in order to use the 741. The essential feature of the 741 is that it is a directly coupled amplifier which has two inputs, an inverting input V_{in-} and a noninverting input V_{in+}. The difference between these two input voltage signals is amplified by a factor of about 10^5 or $100\,\mathrm{dB}$ and the amplified difference voltage appears at the single output terminal. The equation which describes the op-amp is therefore:

$$V_{out} = A_0 \left(V_{in+} - V_{in-} \right)$$

where A_0 is called the open loop gain and $A_0 \approx 10^5$. The complex circuit

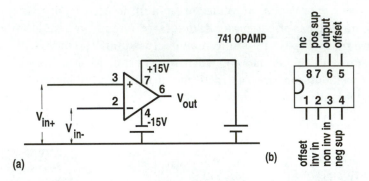

Figure 39.2: Op-amp circuit symbol and pin connections.

shown in Figure 35.5 is therefore replaced by the circuit symbol shown in Figure 39.2.(a). The integrated circuit is powered by $+15\,\mathrm{V}$ and $-15\,\mathrm{V}$ supplies connected with respect to the $0\,\mathrm{V}$ ground line, as shown. The top view of the 8 pin dil package in Figure 39.2 (b) gives the pins associated with each of the op-amp functions. The offset null function on pins 1 and 5 is used for compensating small manufacturing imbalances in the transistors in the input stages of the op-amp. We will examine this function at a later stage.

Now consider the equation for the op-amp gain given above. If we operate the op-amp in the linear region where the output voltage remains within the range $\pm 10\,\mathrm{V}$, the output voltage then corresponds to maximum input voltage differences between the two input terminals of:

$$|V_{in+} - V_{in-}| = \frac{10\,\mathrm{V}}{A_0} = \frac{10}{10^5} = 10^{-4}\,\mathrm{V} = 100\,\mu\mathrm{V}$$

which is small. In typical use the difference of the voltages at the input terminals will usually be much smaller than this so we then obtain the rule that *when an op-amp is used in the linear region, the voltage difference between the inverting and noninverting inputs is approximately zero.* The voltage difference is not exactly zero, otherwise we would never get an output voltage from the op-amp, but it is negligible compared to the voltages normally applied to the amplifier circuit. It is important always to make a distinction between the op-amp and the amplifier circuit which uses an op-amp as a component.

The op-amp has been designed to have an input resistance between the inverting and noninverting inputs of at least $1\,\mathrm{M}\Omega$. We have just seen that the typical voltage difference between the two input terminals is not greater than $100\,\mu\mathrm{V}$ when the op-amp is operating in the linear region. This means that the current flowing into the input terminals is less than $\frac{100\,\mu\mathrm{V}}{10^6\,\Omega} = 10^{-10}\,\mathrm{A}$. This current is so small as to be negligible. We therefore have our second approximation rule: *no current flows into the input terminals of the op-amp.*

These two rules for op-amp operation in the linear region permit nearly all op-amp circuits to be analyzed and give a tremendous simplification of the analysis of op-amp circuits as compared to the problems of analysis of discrete transistor or FET circuits.

We will now examine how these rules can be used to analyze an inverting amplifier such as that shown in Figure 39.3 (a).

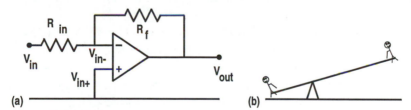

Figure 39.3: Inverting amplifier circuit and model.

The negative power supply connection at pin 4 and the positive power supply connection at pin 7 have been omitted from the diagram but not from the actual circuit in order to avoid cluttering up the diagram. This convention will be followed in the remainder of the text. Another point of note is that the circuit can be drawn with either the inverting input ($-$) or the noninverting input $(+)$ at the top of the op-amp symbol depending on which configuration gives greater clarity and simplicity in the circuit diagram.

Two resistors, R_{in} and R_f, are used in the circuit. A voltage, V_{in}, is applied at the input to the amplifier. We are using the op-amp in the linear region (which implies that we are using negative feedback which we will discuss in Unit 41) and therefore the first rule for op-amps applies — the voltage difference between inverting input and noninverting input is approximately zero.

$$V_{in-} \approx V_{in+} = 0\,\text{V}$$

The second rule tells us that all of the current flowing though R_{in} also flows through R_f because no current flows into the inverting input of the op-amp. This then gives us:

$$I_{in} = I_f$$

but by using Ohm's law we have:

$$\frac{V_{in} - V_{in-}}{R_{in}} = \frac{V_{in} - 0\,\text{V}}{R_{in}} = I_{in}$$

$$\frac{V_{in-} - V_{out}}{R_f} = \frac{0\,\text{V} - V_{out}}{R_f} = I_f$$

$$\text{Therefore} \quad \frac{V_{in}}{R_{in}} = -\frac{V_{out}}{R_f}$$

$$\text{or} \quad A_V = \frac{V_{out}}{V_{in}} = -\frac{R_f}{R_{in}}$$

This gives us an amplifier which has a voltage gain which is determined by the ratio of two resistors. The negative sign tells us that it is an inverting amplifier.

A very good analogy for this circuit is the seesaw shown in Figure 39.3 (b). The pivot point of the seesaw does not move and this is the analog to the first rule. The ratio of the movement of the two children is the inverse of the ratio of the lengths of the two sides of the seesaw. This is the analog to $A_V = -\frac{R_f}{R_{in}}$. One child goes up, the other child goes down. This gives the negative sign corresponding to this inversion of movement.

39.1   Example

39.1 An inverting amplifier powered from a dual $\pm 15\,\text{V}$ supply has $R_{in} = 10\,\text{k}\Omega$ and $R_f = 150\,\text{k}\Omega$. Calculate the gain of the amplifier and plot a graph of the output voltage when V_{in} is varied from $-2\,\text{V}$ to $+2\,\text{V}$.

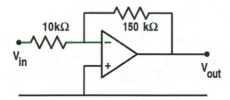

Figure 39.4: Example 39.1. Inverting amplifier.

The gain of this circuit is $A_V = -\frac{150\,\mathrm{k\Omega}}{10\,\mathrm{k\Omega}} = -15$.

Now calculate a number of representative values for the output given by $V_{out} = -15 \times V_{in}$:

V_{in}	Calculated V_{out}	Actual V_{out}
$-2\,\mathrm{V}$	$+30\,\mathrm{V}$	$+13\,\mathrm{V}$
$-1\,\mathrm{V}$	$+15\,\mathrm{V}$	$+13\,\mathrm{V}$
$-0.8\,\mathrm{V}$	$+12\,\mathrm{V}$	$+12\,\mathrm{V}$
$-0.5\,\mathrm{V}$	$+7.5\,\mathrm{V}$	$+7.5\,\mathrm{V}$
$0\,\mathrm{V}$	$0\,\mathrm{V}$	$0\,\mathrm{V}$
$+0.5\,\mathrm{V}$	$-7.5\,\mathrm{V}$	$-7.5\,\mathrm{V}$
$+0.8\,\mathrm{V}$	$-12\,\mathrm{V}$	$-12\,\mathrm{V}$
$+1\,\mathrm{V}$	$-15\,\mathrm{V}$	$-13\,\mathrm{V}$
$+2\,\mathrm{V}$	$-30\,\mathrm{V}$	$-13\,\mathrm{V}$

These calculations are plotted in Figure 39.5. It should be noted that the op-amp goes into saturation for output voltages about 2 V less than the supply voltages which are $\pm15\,\mathrm{V}$ in this case. This gives saturation at $\pm13\,\mathrm{V}$ output.

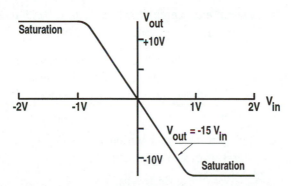

Figure 39.5: Output voltage as a function of input voltage for Example 39.1.

39.2   Problems

39.1 A 10 kΩ potentiometer can be rotated from 0° to 270° and is connected between the +15 V and −15 V supplies. Calculate the output voltage as a function of the angle of rotation. Can the output voltage ever be greater than +15 V or less than −15 V?

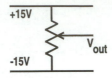

Figure 39.6: Problem 39.1.

39.2 Explain why the output voltage from an amplifier powered from a ±15 V supply can never be greater than +15 V or less than −15 V.

39.3 Calculate the voltage gain of the amplifier shown in Figure 39.7. Plot the output voltage as a function of input voltage for input voltages between −1 V and +1 V. The op-amp is powered from a dual ±12 V supply which is not shown in the circuit diagram.

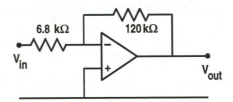

Figure 39.7: Problem 39.3.

39.4 Calculate the current which flows in the 120 kΩ feedback resistor in Figure 39.7 when the input voltage is +0.35 V.

39.5 If the 6.8 kΩ, R_{in} and the 120 kΩ, R_f, in Figure 39.7, are changed to 100 Ω and 2.5 kΩ respectively, calculate the new value of the voltage amplification. Calculate the value of the current in the R_f when the input voltage is 0.15 V.

39.6 Design an inverting amplifier which has a voltage gain of −37. Calculate the input resistance of the amplifier.

39.7 Design an inverting amplifier which has an input resistance of 12 kΩ and a voltage gain of −18.

Unit 40   The noninverting amplifier

- The gain of a noninverting amplifier is given by:

$$A_V = 1 + \frac{R_1}{R_2}$$

- The input resistance of a noninverting amplifier is of the order of $100\,\text{M}\Omega$.

The general form of the circuit used as a noninverting amplifier is shown in Figure 40.1.

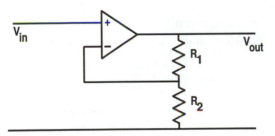

Figure 40.1: The noninverting amplifier circuit.

A fraction of the output voltage, determined by the potential divider at the output of the op-amp, is applied to the inverting input of the op-amp. We then use the first rule for op-amps from Unit 39 to get the result that the voltages at the two inputs to the op-amp are effectively equal. This gives:

$$V_{in} = V_{in+} = V_{in-} = V_{out} \times \frac{R_2}{R_1 + R_2}$$

which is easily simplified to get:

$$A_V = \frac{V_{out}}{V_{in}} = \frac{R_1 + R_2}{R_2} = 1 + \frac{R_1}{R_2}$$

The input resistance of the noninverting amplifier is high because the resistance between the two input terminals is about $1\,\text{M}\Omega$. However, when

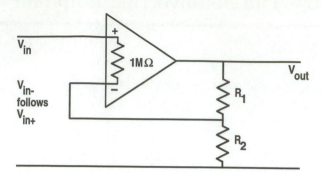

Figure 40.2: Calculating the input resistance.

an input is applied to the noninverting input, the full input voltage does not appear across the $1\,\text{M}\Omega$ since the output voltage is fed back to the other end of the $1\,\text{M}\Omega$, at the inverting input, so as to cause it to follow closely the voltage at the noninverting input. We saw in Unit 39 that the maximum difference between the two inputs is about $100\,\mu\text{V}$. The result is that the amplifier input resistance is much greater than the op-amp input resistance because of the feedback signal. The maximum input current is:

$$I_{in(max)} = \frac{100\,\mu\text{V}}{1\,\text{M}\Omega} = 10^{-10}\,\text{A}$$

which gives a minimum amplifier input resistance of about $10^9\,\Omega$. Various current leakage paths on the printed circuit board due to moisture films, dirt etc., will usually reduce the amplifier input resistance to about $100\,\text{M}\Omega$ unless special precautions are taken to minimize current leakage paths. In most cases an input resistance of $100\,\text{M}\Omega$ is fully adequate and special precautions are unnecessary.

40.1   Example

40.1 The circuit for a noninverting amplifier is shown in Figure 40.3. The power supply voltages are $\pm15\,\text{V}$. Calculate the gain of this amplifier and plot a graph of the output voltage as V_{in} is varied from $-1\,\text{V}$ to $+1\,\text{V}$.

The gain of the amplifier is given by:

$$A_V = 1 + \frac{R_1}{R_2} = 1 + \frac{5000}{200} = 1 + 25 = 26$$

We now calculate a number of representative values for the output voltage using $V_{out} = 26 \times V_{in}$.

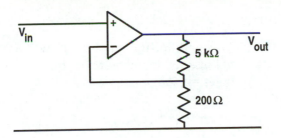

Figure 40.3: Example 40.1.

V_{in}	Calculated V_{out}	Actual V_{out}
$-2\,\text{V}$	$-52\,\text{V}$	$-13\,\text{V}$
$-1\,\text{V}$	$-26\,\text{V}$	$-13\,\text{V}$
$-0.4\,\text{V}$	$-10.4\,\text{V}$	$-10.4\,\text{V}$
$-0.1\,\text{V}$	$-2.6\,\text{V}$	$-2.6\,\text{V}$
$0\,\text{V}$	$0\,\text{V}$	$0\,\text{V}$
$0.1\,\text{V}$	$2.6\,\text{V}$	$2.6\,\text{V}$
$0.4\,\text{V}$	$10.4\,\text{V}$	$10.4\,\text{V}$
$1\,\text{V}$	$26\,\text{V}$	$13\,\text{V}$
$2\,\text{V}$	$52\,\text{V}$	$13\,\text{V}$

These results are plotted in Figure 40.4. Again, as in the case of the
inverting amplifier, it should be noted that the output voltage does not
move outside the limits of $\pm 13\,\text{V}$ when the supply voltage is $\pm 15\,\text{V}$ so
that the output voltage saturates for large values of the input voltage.
The region of amplifier linear response within which $V_{out} = 26 \times V_{in}$ is
indicated on the diagram.

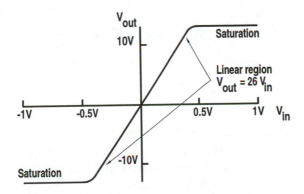

Figure 40.4: V_{out} as a function of V_{in} for Example 40.1.

40.2   Problems

40.1 Calculate the voltage gain for the amplifier shown in Figure 40.5 and plot the output voltage for input voltages from $-1\,$V to $+1\,$V. The power supply voltages for the op-amp are $\pm10\,$V.

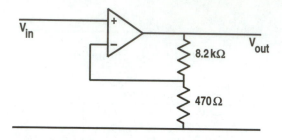

Figure 40.5: Problem 40.1.

40.2 Design a noninverting amplifier which has a gain which can be varied from $+9$ to $+60$. The circuit shown in Figure 40.6 is suggested. Calculate suitable values for the fixed resistors and for the potentiometer resistance, R_V. What other arrangements of resistors could be used in the potential divider?

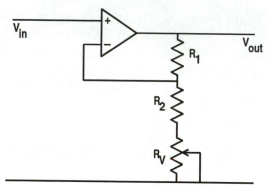

Figure 40.6: Problem 40.2.

40.3 Design a noninverting amplifier which has a gain of $+43$. Draw the full circuit diagram for the amplifier including the power supply connections. The amplifier is to be powered by two PP9, $9\,$V batteries. Show where the batteries should be connected.

Unit 41 Negative feedback

- If a fraction, β, of the output of an amplifier is fed back and subtracted from the input to the amplifier and if the open loop gain of the amplifier is large, the closed loop gain of the amplifier is given by:

$$A_V = \frac{1}{\beta} = \frac{1}{\text{Feedback fraction}}$$

The general configuration of a negative feedback amplifier is shown in Figure 41.1.

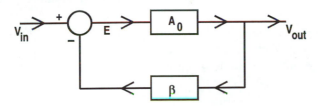

Figure 41.1: Block diagram of negative feedback loop.

The input signal is fed to the +input of a comparator, represented by the circle in the diagram, and the feedback signal is fed to the −input of the comparator. The output of the comparator is the difference of these two signals and is called the error signal, E. (This comparator is part of the op-amp but may be a separate block in other control loops.)

The high gain amplifier, the op-amp in our circuit, amplifies this error signal by an amount A_0 where A_0 is the open loop gain of the amplifier or op-amp. This gives an output:

$$V_{out} = E \times A_0$$

We then take a fraction of this output voltage, β, for instance by using a potential divider, and we feed back a signal $\beta \times V_{out}$ which is subtracted at the −input to the comparator. This then gives the fundamental equation for the negative feedback amplifier:

$$V_{out} = E \times A_0 = (V_{in} - \beta V_{out}) \times A_0$$

which simplifies to give:

$$A_V = \frac{V_{out}}{V_{in}} = \frac{A_0}{1 + A_0\beta} = \frac{1}{\frac{1}{A_0} + \beta}$$

where A_V is the closed loop voltage gain. If the open loop gain is large ($A_0 = 10^5$ for an op-amp) then $\frac{1}{A_0}$ is very small compared to β and the gain of the negative feed back amplifier is given by:

$$A_V \approx \frac{1}{\beta} = \frac{1}{\text{Feedback fraction}}$$

This allows the gain of the amplifier, A_V, to be determined to a high degree of accuracy by passive components such as resistors.

Negative feedback systems are used in many areas of modern industry as well as in electronic circuits. The applications may in general be divided into two groups:

Control systems. A typical example of a control system is a chemical process where a uniform product is to be delivered even though the load or process throughput may vary. Figure 41.2 shows the general configuration of such a control system.

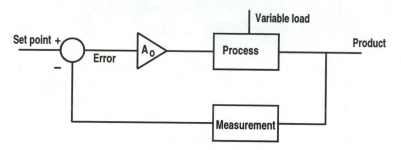

Figure 41.2: Generalized control loop.

The characteristic feature of a control loop is that the set point is fixed but that the load or throughput may vary. The loop maintains the product operating point (temperature, pressure etc.) constant. The table shows, on each horizontal row, some typical applications of negative feedback control systems and some of the associated variables within the control loop.

Application	Set point	Measurement	Control method	Load
DC power supply	Voltage	Voltmeter	Transistor	Current drawn
Hot water heater	Temperature	Thermometer	Heater	Water flow
Traffic	Speed limit	Speedometer	Accelerator	Road incline
Economy	Growth	Inflation	Interest rates	Rest of world

In general, negative feedback control systems act in such a way as to minimize the error, in spite of external factors, so as to obtain a uniform product and a stable system.

In contrast, a good example of the instabilities associated with a positive feedback system are the Black Mondays associated with the effects of the computerized share dealing programs used by dealers on the stock exchange. These programs introduce positive feedback by selling on a falling market and buying on a rising market.

Servo systems comprise the second main group of negative feedback applications. These systems are illustrated in Figure 41.3.

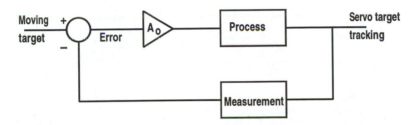

Figure 41.3: Generalized servo loop.

In this group the target or set point varies and the servo loop acts to force the process to follow the varying target. The load in general is constant or nearly so.

A good example of a servo system would be an automatic guided vehicle which follows a white, painted track around a factory floor as it delivers parts to the various machines.

Another example would be a national economy, viewed from the point of view of a government coming up for re-election, in which the apparent growth rate is manipulated by controlling interest rates at the expense of inflation.

41.1   Example

41.1 Analyze the noninverting amplifier shown in Figure 41.4 in terms of a negative feedback system.

In this case the fraction of the output that is fed back is determined by the potential divider on the output voltage.

The feedback fraction $\beta = \dfrac{R_2}{R_1 + R_2}$

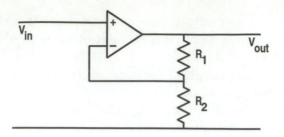

Figure 41.4: Example 41.1.

Therefore the gain of the amplifier is:

$$
\begin{aligned}
A_V &= \frac{1}{\beta} \\
&= \frac{1}{\frac{R_2}{R_1 + R_2}} \\
&= \frac{R_1 + R_2}{R_2} \\
&= 1 + \frac{R_1}{R_2}
\end{aligned}
$$

which is the same result which we obtained in Unit 40 using a different method.

41.2   Problems

41.1 List four different negative feedback systems with which you are familiar. In each example, identify the controlled variable, the measuring element and the control element.

41.2 The op-amp used in the amplifier shown in Figure 41.5 has an open loop gain, $A_0 = 10^5$. Use the basic feedback equation:

$$ V_{out} = A_0 \left(V_{in} - \beta V_{out} \right) $$

to determine the exact voltages at each of the three signal terminals of the op-amp when the input signal, $V_{in} = 0.10 \, \text{V}$.

What is the output voltage calculated, using the formula for the gain, $A_V = 1 + \frac{R_1}{R_2}$, derived in Unit 40? Is there any significant difference between the two answers?

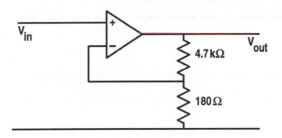

Figure 41.5: Problem 41.2.

41.3 If the op-amp used in Problem 41.2 is replaced by a low specification op-amp having an open loop gain, $A_0 = 5 \times 10^3$, calculate the new output voltage for the same input voltage of 0.10 V. Does the use of a low gain amplifier cause a significant change in the output voltage?

41.4 Design a noninverting amplifier which has a gain of +75. Calculate suitable component values and draw the full circuit including the power supply connections to the op-amp. What is the value of the feedback fraction for the amplifier?

41.5 A negative feedback controller has a response curve as shown in Figure 41.6 where the output of the controller is plotted against the measurement expressed as a percentage of the range. If the error is zero when the system is operating at 50% of load, calculate the error for 10% of load and for 60% of load.

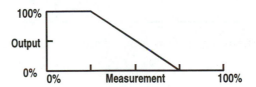

Figure 41.6: Problem 41.5.

Unit 42 Adder circuits

- The output from a two input noninverting adder is given by:

$$V_{out} = V_1 + V_2$$

- The output from an inverting adder is given by:

$$V_{out} = -R_f \left(\frac{V_1}{R_1} + \frac{V_2}{R_2} + \frac{V_3}{R_3} + \dots \right)$$

The addition of two voltage signals to get a single output sum is a very common operation in electronics. We will first look at a noninverting simple adder as it is a good example of circuit analysis techniques where a given circuit can be split into its component parts in order to analyze the operation of the circuit. The circuit is shown in Figure 42.1.

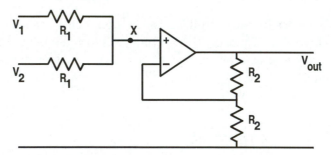

Figure 42.1: Two input noninverting adder.

The two input voltages, V_1 and V_2, are specified with respect to ground and are applied to the ends of two equal resistors of value R_1, connected in series. Since no current flows into the input terminals of the op-amp, current only flows in the resistors. The current in the two R_1 in series is given by:

$$I = \frac{V_1 - V_2}{2 \times R_1}$$

The voltage, V_X, at the mid point, X, of the resistors is given by:

$$V_X = V_1 - I \times R_1 = V_1 - \frac{V_1 - V_2}{2 \times R_1} \times R_1 = V_1 - \frac{V_1}{2} + \frac{V_2}{2} = \frac{V_1 + V_2}{2}$$

In other words, V_X is the average of V_1 and V_2.

The noninverting amplifier has a gain of $A_V = 1 + \frac{R_2}{R_2} = 2$. The output of the circuit is then given by:

$$V_{out} = \frac{V_1 + V_2}{2} \times 2 = V_1 + V_2$$

which is the sum of the two input voltages. The disadvantage of this adder is that no more than two input signals can be added.

The inverting adder is a much more versatile type of adder circuit and a typical circuit configuration is shown in Figure 42.2.

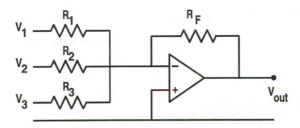

Figure 42.2: Inverting adder.

This particular circuit has three input signals but any number of inputs can be used.

The key to the analysis of this circuit is the first rule for op-amps (Unit 39) that the voltage at the inverting input is the same as the voltage at the noninverting input. In this case the noninverting, (+), input is at zero volts because it is connected to ground. Therefore the voltage at the inverting, (−), input is at approximately zero volts. This is called a *virtual earth* or *virtual ground*.

Now calculate the current which flows in each of the input resistors. The voltage difference across R_1 is $V_1 - 0 = V_1$ and the current is therefore $\frac{V_1}{R_1}$. A similar calculation is carried out for each of the other inputs.

The voltage across the feedback resistor, R_f, is $0 - V_{out} = -V_{out}$ and therefore the feedback current is $I_f = \frac{-V_{out}}{R_f}$. Since no current flows in the input terminal of the op-amp (Rule 2, Unit 39) we get $I_1 + I_2 + I_3 = I_f$ and therefore:

$$\frac{V_1}{R_1} + \frac{V_2}{R_2} + \frac{V_3}{R_3} = -\frac{V_{out}}{R_f}$$

The advantage of this inverting adder is that it allows us to carry out a scaled or weighted addition to get an output of the form:

$$V_{out} = -(3.9V_1 + 4.0V_2 + 1.0V_3)$$

The inversion is easily reversed by using an inverter circuit having a gain of -1.

If an output of the form:

$$V_{out} = 3.1V_1 - 2.4V_2 - 5.0V_3$$

is required, then an inverter can be used on the V_1 input signal before it is applied to the inverting adder so that the effective equation is:

$$V_{out} = -\left(3.1(-V_1) + 2.4V_2 + 5.0V_3\right)$$

42.1   Examples

42.1 Derive the equation for the output voltage from the circuit shown in Figure 42.3.

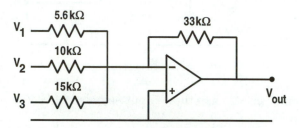

Figure 42.3: Example 42.1.

The equation for the output is:

$$
\begin{aligned}
V_{out} &= -\left(\frac{33}{5.6} \times V_1 + \frac{33}{10} \times V_2 + \frac{33}{15} \times V_3\right) \\
&= -5.9V_1 - 3.3V_2 - 2.2V_3
\end{aligned}
$$

42.2 A circuit such as that in Figure 42.3 has been constructed. Design a systematic test procedure for the circuit.

There are three inputs to the circuit so each of these three input channels must be tested separately and the measured output compared with the calculated output. Possible test voltages are zero volts, and small positive and negative voltages. Only after the individual channels have been tested should test signals be applied to two or more channels.

A test matrix such as that shown below should be prepared and the specified measurements carried out and filled into the column marked

Measured V_{out}. It is only by carrying out systematic testing that you can ever be sure that a circuit you design and construct actually works to specification.

The equation describing the circuit to be tested is:

$$V_{out} = -5.9V_1 - 3.3V_2 - 2.2V_3$$

Test no.	V_1	V_2	V_3	Calculated V_{out}	Measured V_{out}
1	0	0	0	0	
2	1	0	0	−5.9	
3	0	1	0	−3.3	
4	0	0	1	−2.2	
5	−1	0	0	+5.9	
6	0	−1	0	+3.3	
7	0	0	−1	+2.2	
8	−1	1	1	+0.4	
9	1	−1	1	−4.8	

42.2 Problems

42.1 Explain the function of each of the tests in the test matrix in Example 42.2. What is being tested and what section of the circuit would you examine in the event of a discrepancy between the Calculated V_{out} and the Measured V_{out}?

42.2 Apply the principle of superposition (Unit 21) to analyze the noninverting adder shown in Figure 42.1.

42.3 Design an adder circuit which will have an output given by:

$$V_{out} = -(2.8V_1 + 3.2V_2 + 1.0V_3 + 1.9V_4)$$

Design a test matrix, similar to that in Example 42.2, for this circuit and calculate the expected outputs for each test input.

42.4 The output voltage from an inverting adder circuit is given by:

$$V_{out} = -1.3V_1 - 2.9V_2$$

If the signals V_1 and V_2 are $V_1 = 2.1\sin(400t)$ V and $V_2 = 0.6$ V, where t is the time in seconds, give a scaled sketch of the output voltage waveform from the adder.

Unit 43   Sensors and interfacing

- The output from a current to voltage converter is given by:

$$V_{out} = -I \times R_f$$

- A bridge is said to be in balance when:

$$\frac{R_1}{R_2} = \frac{R_3}{R_4}$$

- The output from a differential amplifier is given by:

$$V_{out} = \frac{R_2}{R_1} \times (V_2 - V_1)$$

A transducer or sensor can be described as a device which responds to an external environment or stimulus and gives an output signal which is a function of one parameter of the environment or stimulus. An example might be a pressure sensor which gives an output voltage proportional to the pressure in a container or a platinum metal resistor whose resistance increases in proportion to the temperature.

While some commercial transducers contain embedded electronics which give output voltages proportional to the parameter being sensed, the signals from many sensors need conditioning or amplification or conversion to voltage signals before the signals can be used. In this unit we examine some common sensors and common signal conditioning techniques.

Photodiodes and current to voltage conversion. Photodiodes or solar cells are large area silicon pn junction diodes, mounted on supporting substrates and encapsulated in housings which allow light to fall on the pn junction region. The light is absorbed in the silicon and generates electron-hole pairs. In the dark, the photodiode behaves just like a normal pn diode but when light falls on the device the electron-hole pairs generated give minority carriers and a significant reverse bias diode current. The characteristics of the device are shown in Figure 43.1.

In the top right and the bottom left quadrant of operation, an external voltage drives current through the device. When the diode is in reverse

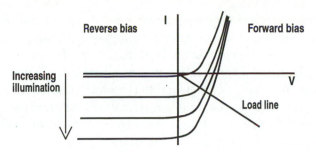

Figure 43.1: Photoconductive and photovoltaic response of photodiode.

bias, the current is proportional to the intensity of illumination. In the bottom right hand quadrant, the signs of the current and the voltage are opposite. This implies that the device can drive current through an external load which is represented by the load line in Figure 43.1. This is the quadrant of operation of the solar cells used to power satellites and also used to power electrical equipment in isolated locations.

We wish to use the photodiode as a light detector, so we can use it in the reverse bias mode and then we get a current which is proportional to the light falling on the device or we can use it in the forward bias mode and then we measure the output voltage from the device. In this second configuration, the output voltage is not proportional to the intensity of the light.

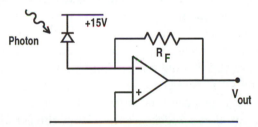

Figure 43.2: Photodiode in photoconductive mode used with a current to voltage converter.

Figure 43.2 shows a circuit which applies a reverse bias to the photodiode so that the diode cathode is at +15 V. The anode is at 0 V (because of Rule 1 of Unit 39). The photodiode current, I_{pd}, also flows in R_f. The voltage across R_f is $0\,\text{V} - V_{out}$ and therefore we get:

$$V_{out} = -I_{pd} \times R_f$$

so we have an output voltage which is proportional to transducer current with the scaling determined by the feedback resistor, R_f. This circuit configuration is called a current to voltage converter or *I–V* converter. The current through a reverse biased photodiode is proportional to the illumination of the photodiode and this circuit then gives an output voltage which is proportional to the intensity of the light falling on the photodiode.

Bridge circuits. There are many sensors whose response to a stimulus takes the form of a change in resistance. However, many of these sensors also show some sensitivity to other ambient stimuli. A common method of dealing with this cross sensitivity is to use a bridge system of two resistive sensors in series. One sensor is exposed to the stimulus to be measured, the other dummy sensor is either made insensitive to the stimulus or shielded from the stimulus. Both sensors are exposed to the environment. Any environmental effects balance out but the stimulus to be measured remains.

A flammable gas sensor for detecting hydrogen or methane in the atmosphere is a good example. A current is passed through a palladium coated filament, heating the filament. If the heated filament is exposed to a low concentration of flammable gas, catalytic combustion occurs on the filament, raising the filament temperature and causing the resistance of the filament to increase. A single coated filament would not give a satisfactory gas sensor since the filament temperature and resistance also change with ambient temperature, power supply variations, ambient humidity and local air flow. The small effect due to the flammable gases would not be resolvable and spurious alarms would result.

The solution is to use a dummy filament, which is not coated with palladium and therefore does not show a temperature rise and resistance increase due to catalytic combustion of flammable gas. The two filaments show similar temperature and resistance changes due to fluctuations in supply voltage, ambient temperature and cooling due to ambient air flow. The two filaments are mounted in the same metal gauze covered flameproof housing and are wired in series in a bridge circuit as shown in Figure 43.3. The gauze covering serves to cool any flame which occurs within the sensor due to ignition of gases within the housing and prevent the flame from escaping and causing ignition of the gases external to the sensor. The gauze permits a free diffusion of ambient gases into the sensing region.

The voltage at the centre of the two filaments is compared to the voltage at the centre of the potential divider of two $1\,\text{k}\Omega$ resistors. If there is no flammable gas present then the differential bridge output voltage, marked V_{out}, is zero. This should be true even if the sensor supply voltage of $3\,\text{V}$ drifts from its nominal value. The output voltage should also be zero even if the ambient temperature changes or there is a moderate air flow near the filaments because the two filaments are affected equally and the bridge remains in balance. However, a small concentration of flammable gas changes the sensing filament but not the dummy filament and the bridge then gives

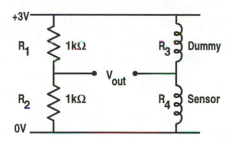

Figure 43.3: Bridge circuit for flammable gas detector.

a nonzero output. The output of such a bridge is usually very small and the signal requires amplification. The bridge output signal is usually amplified with a differential amplifier or difference amplifier.

The differential amplifier. A typical application of the differential amplifier is to measure small voltage difference signals which occur in circuits such as the bridge circuit. In a bridge circuit there is a usually large common mode signal, about $1.5\,\mathrm{V}$ in the case of the flammable gas sensor bridge in Figure 43.3, which is applied to both inputs. A small difference voltage is to be measured in the presence of this large common mode signal. The circuit which is usually used is the differential amplifier such as that shown in Figure 43.4 (a). To analyze the operation of this differential amplifier circuit, we

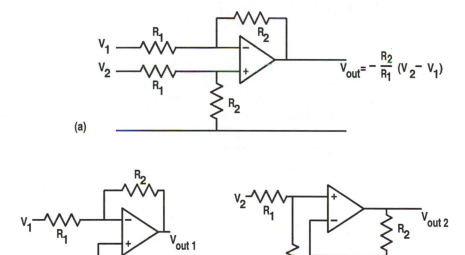

Figure 43.4: The differential amplifier.

apply the principle of superposition (Unit 21).

Set $V_2 = 0$ and short this input. The circuit then becomes an inverting amplifier as shown in Figure 43.4 (b) which has an output due to input signal V_1 of:

$$V_{out1} = -\frac{R_2}{R_1} \times V_1$$

Now find the response due to input V_2 alone by shorting V_1 input to ground. This gives the circuit in Figure 43.4 (c). The circuit is the same as Figure 43.4 (a) but has been redrawn with a different layout while maintaining the same topology or interconnections. It can be seen that this circuit is really a potential divider of R_1 and R_2 which gives a signal of $\frac{R_2}{R_1+R_2} \times V_2$ at the +input to the noninverting amplifier configuration. The noninverting amplifier has a gain of:

$$A_V = 1 + \frac{R_2}{R_1} = \frac{R_1 + R_2}{R_1}$$

The output signal due to input signal V_2 acting alone is therefore:

$$V_{out2} = \frac{R_2}{R_1 + R_2} \times \frac{R_1 + R_2}{R_1} \times V_2 = \frac{R_2}{R_1} \times V_2$$

We then use the principle of superposition to obtain the response when both input signals V_1 and V_2 are present and get an output from the differential amplifier of:

$$V_{out} = \frac{R_2}{R_1} \times (V_2 - V_1)$$

Self balancing bridge. If one resistor of the four in a bridge is a sensing element, as shown in Figure 43.5, for which the resistance is given by an

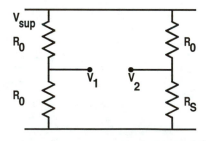

Figure 43.5: Out of balance bridge.

equation of the form $R_S = R_0(1 + \alpha)$, then it is found that the differential bridge voltage is given by:

$$V_{Bridge} = V_1 - V_2 = \frac{V_{sup}}{2} - \frac{R_S}{R_0 + R_S} \times V_{sup}$$

$$= V_{sup} \left(\frac{1}{2} - \frac{R_0(1 + \alpha)}{2R_0 + \alpha R_0} \right)$$

which is a nonlinear function of α.

This means that a linear resistive sensor, when used in a fixed or out of balance bridge circuit, gives a nonlinear output voltage response.

A very useful way of maintaining linearity is to use a self balancing bridge where the op-amp acts in such a way as to restore the bridge to balance. A suitable circuit is shown in Figure 43.6 (a). The bridge is in balance and the output from the op-amp $V_{out} = 0$ when $\alpha = 0$ in $R_S = R_0(1 + \alpha)$.

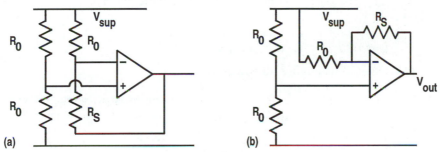

Figure 43.6: Self balancing bridge circuit.

This circuit is best analyzed by redrawing the circuit with a different topology as shown in Figure 43.6 (b). The input to the noninverting, +input, is now $\frac{V_{sup}}{2}$. This gives a fixed offset voltage, V_{off}. The input to the inverting amplifier comprising the op-amp, R_0 and R_S is:

$$V_{sup} - V_{off} = V_{sup} - \frac{V_{sup}}{2} = \frac{V_{sup}}{2}$$

The gain of the inverting amplifier is $-\frac{R_S}{R_0}$. So we get the output voltage of:

$$V_{out} = -\frac{R_S}{R_0} \left(\frac{V_{sup}}{2} \right) = -(1 + \alpha) \times \frac{V_{sup}}{2}$$

which is a linear function of α.

This bridge is called a self balancing bridge because the action of the op-amp maintains zero voltage difference between the two mid points of the bridge by varying the voltage applied across R_S.

43.1   Problems

43.1 The photodiode in the circuit shown in Figure 43.7 has a reverse current sensitivity to incident light of $0.3\,\mathrm{A\,W^{-1}}$ and has a sensitive area of $2.5\,\mathrm{mm^2}$.

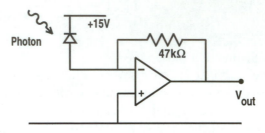

Figure 43.7: Problem 43.1.

Calculate the output voltages from the circuit when the photodiode is exposed to:

(a) Daylight having an intensity of $200\,\mathrm{W\,m^{-2}}$.

(b) Dusk light having an intensity of $5\,\mathrm{W\,m^{-2}}$.

43.2 A thermistor is a metal oxide semiconductor resistor whose resistance decreases with increasing temperature. The resistance of the MA473 thermistor used in the circuit shown in Figure 43.8 is given by:

$$R = 47\,\mathrm{k\Omega} \times \exp\left(\frac{3940}{T} - \frac{3940}{298}\right)$$

where T is the temperature in K.
Calculate the output voltages from the circuit for thermistor temperatures of $0\,^\circ\mathrm{C}$, $30\,^\circ\mathrm{C}$, $45\,^\circ\mathrm{C}$ and $75\,^\circ\mathrm{C}$.

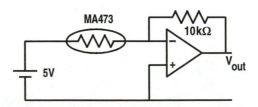

Figure 43.8: Thermistor thermometer.

43.3 The PRT_{100} is an industry standard platinum resistance thermometer having a resistance of $100\,\Omega$ at $0°C$. The temperature coefficient of resistance of platinum metal is such that the resistance of a PRT_{100} varies linearly with temperature and is $138.5\,\Omega$ at $100°C$. A PRT_{100} is used in the bridge circuit shown in Figure 43.9.

Calculate the out of balance signal at temperatures of $20°C$, $35°C$, $75°C$ and $100°C$.

If a differential amplifier, such as that in Figure 43.4 (a) and having a voltage gain of $+10$, is connected across the bridge, calculate the amplifier output at these four values of temperature.

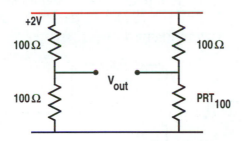

Figure 43.9: Problem 43.3. PRT_{100} temperature bridge.

43.4 A linear Hall effect IC, used for measuring magnetic fields, is shown in Figure 43.10. The device gives a differential output voltage between pins 2 and 3 when placed in a magnetic field. The sensitivity of the sensor is $8.2\,V\,T^{-1}$ where the magnetic field is in tesla (T). Calculate the output voltage from the circuit in Figure 43.10 when the Hall effect IC is (a) in the magnetic field of the Earth ($\approx 6 \times 10^{-5}\,T$) and (b) in the magnetic field of a permanent magnet ($0.05\,T$).

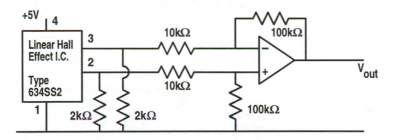

Figure 43.10: Problem 43.4. Hall effect probe.

Unit 44 Differentiator circuits

- The output from a differentiator circuit is given by:

$$V_{out} = -CR_f \frac{dV_{in}}{dt}$$

When we feed a signal into a differentiator circuit, we obtain an output voltage which is the time rate of change of the input voltage signal. The basic circuit used is shown in Figure 44.1.

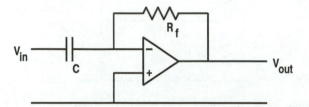

Figure 44.1: The differentiator circuit.

This circuit is analyzed by again using the two rules from Unit 39. The voltage across the capacitor gives the charge on the capacitor as:

$$Q = C \times (V_{in} - 0) = C \times V_{in}$$

The input current is the time rate of change of charge which gives:

$$I_{in} = \frac{dQ}{dt} = C \frac{dV_{in}}{dt}$$

Since no current flows into the op-amp terminals, the same current must also flow in the feedback resistor, R_f, which gives:

$$I_{in} = I_f = \frac{0 - V_{out}}{R_f} = -\frac{V_{out}}{R_f}$$

resulting in the basic equation for differentiators:

$$V_{out} = -CR_f \frac{dV_{in}}{dt}$$

44.1 Example

44.1 In the circuit in Figure 44.2 (a), the function generator, FG, is set to give an output triangular waveform of frequency 1 kHz and amplitude 10 mV as shown in Figure 44.2 (b). Calculate the output voltage waveform and sketch the waveform which would be observed on a two channel oscilloscope with channel A displaying the input signal and channel B displaying the output signal.

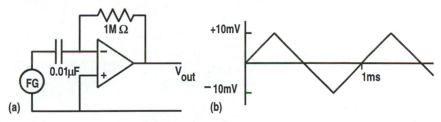

Figure 44.2: Example 44.1.

First calculate the time constant, T, in seconds:

$$T = CR = 0.01\,\mu\text{F} \times 1\,\text{M}\Omega = 10^{-2} \times 10^{-6} \times 10^6 = 10^{-2}\,\text{s}$$

The period of the waveform is $P = \frac{1}{f} = \frac{1}{1000} = 1\,\text{ms}$

The input waveform goes from $\pm 10\,\text{mV}$ to $\mp 10\,\text{mV}$ in 0.5 ms and therefore the rate of change of the input signal is:

$$\left|\frac{dV_{in}}{dt}\right| = \frac{10\,\text{mV} - (-10\,\text{mV})}{0.5 \times 10^{-3}\,\text{s}} = \pm 40\,\text{V s}^{-1}$$

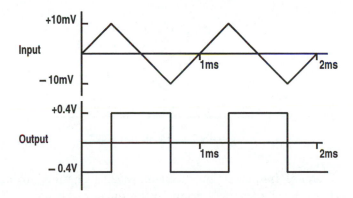

Figure 44.3: Oscilloscope trace of input and output for Example 44.1.

The output voltage is therefore:

$$V_{out} = \pm 10^{-2}\,\text{s} \times 40\,\text{V s}^{-1} = \pm 0.4\,\text{V}$$

The oscilloscope display is shown in Figure 44.3. You should pay particular attention to the alignment of the two waveforms and to the sign of the output square wave signal. When the input is increasing, the output is negative due to the inverting amplifier configuration.

44.2   Problems

44.1 Calculate the output voltage waveform from the circuit in Figure 44.4 for an input triangular waveform of frequency 250 Hz and of amplitude 30 mV.

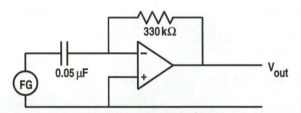

Figure 44.4: Problem 44.1.

44.2 Calculate the output voltage waveform from the circuit in Figure 44.5 when the function generator is set to give an input sinusoidal waveform of frequency 200 Hz and of amplitude 0.1 V. Sketch the traces for the input and output voltage waveforms which you would observe on a double channel oscilloscope.

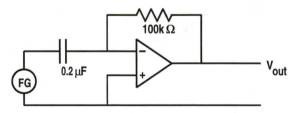

Figure 44.5: Problem 44.2.

44.3 If the signal frequency in Problem 44.2 is changed from 200 Hz to 400 Hz, what changes will occur in the output voltage waveform?

Unit 45   Integrator circuits

- The output from an integrator circuit is given by:

$$V_{out} = -\frac{1}{CR} \int V_{in} dt$$

When we feed a signal into an integrator circuit, we obtain an output voltage which is the time integral of the input voltage signal. The basic circuit used is shown in Figure 45.1.

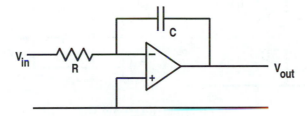

Figure 45.1: The integrator circuit.

This integrator circuit is analyzed by using the same equation for a capacitor, $Q = C \times V$, that we used in analyzing the differentiator circuit. The sequence of the argument is reversed, however.

$$\text{We have, for the input current,} \quad I_{in} = \frac{V_{in}}{R}$$

$$\text{also} \quad Q = \int I_{in} dt$$

$$\text{Therefore the output voltage} \quad V_{out} = -\frac{Q}{C}$$

$$= -\frac{1}{C} \int_0^t I_{in} dt$$

$$= -\frac{1}{C} \int_0^t \frac{V_{in}}{R} dt$$

$$= -\frac{1}{CR} \int_0^t V_{in} dt$$

225

which means that the output voltage from this circuit is the time integral of the input voltage to the circuit.

45.1   Example

45.1 In the circuit in Figure 45.2 (a), the function generator, FG, is set to give an output square waveform of frequency 200 Hz and of amplitude 1 V as shown in Figure 45.2 (b).

Calculate the output voltage waveform and sketch the waveform which would be observed on a two channel oscilloscope with channel A displaying the input waveform and channel B displaying the output waveform.

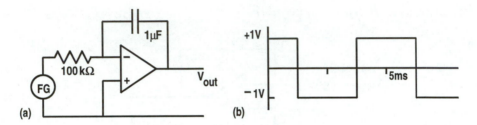

Figure 45.2: Example 45.1.

First calculate:

$$CR = 1\,\mu\text{F} \times 100\,\text{k}\Omega = 10^{-6} \times 10^{5} = 0.1\,\text{s}$$

We then use the equation for the integrator to get:

$$
\begin{aligned}
V_{out} &= -\frac{1}{CR}\int_0^t V_{in}dt \\
&= -\frac{1}{0.1}\int_0^t V_{in}dt \\
&= -10\int_0^t V_{in}dt
\end{aligned}
$$

V_{in} is constant for 1.25 ms at +1 V and then
V_{in} is constant for 2.5 ms at −1 V and then
V_{in} is constant for 2.5 ms at +1 V and then
V_{in} is constant for 2.5 ms at −1 V and then ... etc.
Also we presume that $V_{out} = 0$ V at the start of the time interval.

At the end of the first $\frac{1}{4}$ segment of the waveform at $t = 1.25\,\text{ms}$:
$V_{out} = -10 \times 1\,\text{V} \times 1.25 \times 10^{-3} = -12.5\,\text{mV}$
During the next $2.5\,\text{ms}$, the output changes by:
$\Delta V_{out} = -10 \times (-1\,\text{V}) \times 2.5 \times 10^{-3} = +25\,\text{mV}$
During the next $2.5\,\text{ms}$, the output changes by:
$\Delta V_{out} = -10 \times (1\,\text{V}) \times 2.5 \times 10^{-3} = -25\,\text{mV}$
which gives the waveform shown in Figure 45.3.

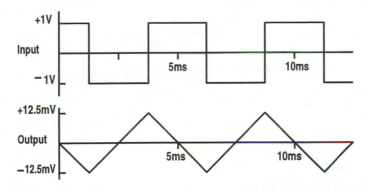

Figure 45.3: Oscilloscope trace of input and output for Example 45.1.

You should note the inversion:
when the input is greater than $0\,\text{V}$, the output decreases,
when the input is less than $0\,\text{V}$, the output increases.

45.2 Problems

45.1 The two switches in the circuit in Figure 45.4 are opened at time $t = 0\,\text{s}$. Calculate the resulting output voltage as a function of time for time from $t = 0\,\text{s}$ to $t = 30\,\text{s}$. Do the calculations accurately describe the variation of the actual output voltage? If not, why not?

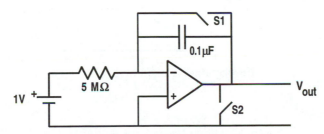

Figure 45.4: Problem 45.1.

45.2 What would be the change in the output voltage waveform in Example 45.1, if the first 1.25 ms at +1 V were omitted from the input signal?

45.3 A sinusoidal voltage waveform having an amplitude of 7.3 V and frequency 20 Hz is applied to the input of the integrator circuit shown in Figure 45.5. Calculate the output voltage waveform and sketch the input and output waveforms which you would observe on a double channel oscilloscope.

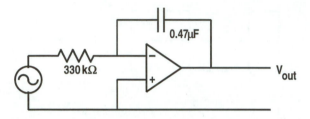

Figure 45.5: Problem 45.3.

45.4 If the frequency of the signal in Problem 45.3 is increased from 20 Hz to 500 Hz, what is the change in the amplitude of the output voltage waveform?

45.5 Does the amplitude of the output signal from an integrator increase or decrease when the input signal frequency is increased?
Two sinusoidal signals of amplitude 1.0 V and frequency 60 Hz and 2.9 V and frequency 170 Hz are applied to the inputs to the circuit shown in Figure 45.6. Use the principle of superposition to calculate the output waveform and give a sketch of the input and output waveforms.

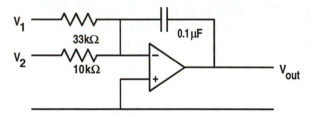

Figure 45.6: Problem 45.5.

Unit 46 Functional amplifiers

If the I–V characteristic for a device is described by $I = f(V)$ then:

- putting the device in place of the input resistor of an inverting amplifier gives the forward function:

$$V_{out} = -R \times f(V_{in})$$

- putting the device in place of the feedback resistor of an inverting amplifier gives the inverse function:

$$V_{out} = -f^{-1}\left(\frac{V_{in}}{R}\right)$$

Suppose we have a two terminal device or component which has an I–V characteristic which is described by a monotonic function $I = f(V)$.

The device can be used in place of the input resistor in an inverting amplifier as shown in Figure 46.1.

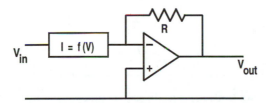

Figure 46.1: The forward function.

Then, from Unit 39, we have the relationship that the current in the input device is equal to the current in the feedback device, $I_{in} = I_f$, and therefore:

$$V_{out} = -R \times I_f = -R \times I_{in} = -R \times f(V_{in})$$

so that we can now generate an arbitrary function of an input voltage if we have available a device with the appropriate current-voltage characteristic.

If the device is used in the feedback path, then the inverse function is generated. This circuit is shown in Figure 46.2.

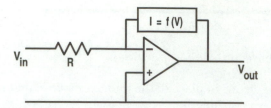

Figure 46.2: The inverse function.

Again the currents in the input and feedback paths are equal so:

$$\frac{V_{in}}{R} = I_{in} = I_f = f\left(-V_{out}\right)$$

which, if f^{-1} represents the inverse function of f, immediately gives:

$$V_{out} = -f^{-1}\left(\frac{V_{in}}{R}\right)$$

46.1   Example

46.1 Derive an expression for the output voltage from the circuit shown in Figure 46.3.

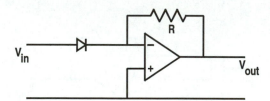

Figure 46.3: Example 46.1. Antilog function amplifier.

In this case we have a diode for which the approximate I–V characteristic in forward bias is:

$$I = I_0 \exp\left(\frac{V}{25\,\text{mV}}\right)$$

Since the device is in the input path we use:

$$V_{out} = -R \times f(V_{in}) = -R \times I_0 \exp\left(\frac{V_{in}}{25\,\text{mV}}\right)$$

which, if we combine the constants into k, gives:

$$V_{out} = -k\log^{-1}\left(V_{in}\right) \quad \text{or} \quad -k\,\text{antilog}(V_{in})$$

The accuracy of this antilog function depends on the accuracy of the exponential function for the diode. Special diodes, called logging diodes, are available which obey the exponential function over about seven decades but it is found that a diode connected transistor in which the base is connected to the collector gives a very good log response.

46.2 Problems

46.1 Describe the behaviour of a functional amplifier which uses a device having a nonmonotonic function characteristic in the input side of the functional amplifier.

46.2 Give an approximately scaled sketch of the output voltage from the circuit in Figure 46.4 as a function of the input voltage for the range $-1\,\text{V} \le V_{in} \le +1\,\text{V}$. Suggest a reason for including the $470\,\Omega$ resistor in series with the input.

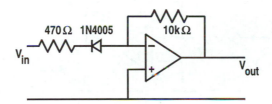

Figure 46.4: Problem 46.2. Antilog function amplifier.

46.3 Give an approximately scaled sketch of the output voltage from the circuit in Figure 46.5 as a function of the input voltage for the range $0\,\text{V} \ge V_{in} \ge +5\,\text{V}$.

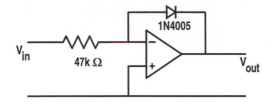

Figure 46.5: Problem 46.3. Log function amplifier.

46.4 Calculate the output as a function of the input for the circuit shown in Figure 46.6. The transistor in this circuit is connected as a transdiode and acts as a high accuracy log response diode. Show that the range of the input voltage is compressed to a small range of output voltage.

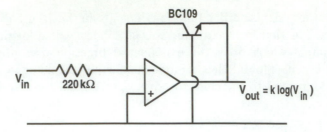

Figure 46.6: Problem 46.4. Log function using transdiode.

46.5 Use the relationship $V_{out} = -f^{-1}\left(\frac{V_{in}}{R}\right)$ to show that the output from the circuit in Figure 46.7 is given by $V_{out} = -\frac{1}{RC}\int V_{in}dt$.

Calculate the amplitude of the output voltage waveform if the input sinusoidal signal is at a frequency of 800 Hz and has an amplitude of 2 V. Note that for a capacitor $I = C\frac{dV}{dt}$.

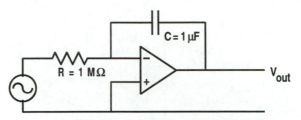

Figure 46.7: Problem 46.5.

46.6 What is the characteristic I-V function for an inductor? A sinusoidal signal of 1 kHz and amplitude 2 V is applied at the input to the circuit in Figure 46.8. Derive the function which describes the output voltage waveform and calculate the amplitude of the output voltage.

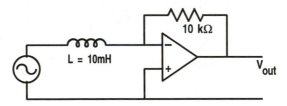

Figure 46.8: Problem 46.6.

46.7 The input signal to an amplifier is applied to the X input of an oscilloscope operating in XY mode and the amplifier output is applied to the oscilloscope Y input. Show that this configuration causes the response curves for Problems 46.2 to 46.6 to be displayed on the screen.

Unit 47 Analog computers

- In analog computers, quantities or variables are represented by voltages.

- Computations are carried out by passing the voltage signals through adders, inverters, differentiators and integrators and other functional amplifiers.

- The results of individual computations are combined by using adders.

We have already met most of the circuit blocks which are used in analog computers. These are DC amplifiers, adders, differentiators, integrators and logarithmic amplifiers.

There are some other circuit blocks which are used in analog computers but which we have not examined. These are multipliers, dividers and square root extractors all of which are readily available as standard ICs for which data sheets are available. We will leave these for other courses.

In analog computers, voltages are used to represent physical real world quantities and the circuit blocks are used to carry out computations on these voltages. In contrast, in digital computers, numbers are used to represent physical real world quantities and the numbers are then manipulated digitally in accordance with a program. In analog computing, the circuit can be considered to be the program, which means that any revision of the program requires a circuit redesign.

Analog computers are often used because of the low cost of op-amps, the robustness of analog control systems and because many real world sensors have voltage outputs which are directly compatible with the inputs of analog computing systems.

We have already seen an example of analog computing in Example 42.1 which showed how a number of voltage inputs could be scaled and added.

47.1 Example

47.1 Design a circuit which implements the following function of the input voltage, V_{in}.

$$V_{out} = 3.3V_{in} + 2.2\frac{dV_{in}}{dt}$$

233

This function represents a controller having gain 3.3 and derivative or rate action of $2.2\frac{dV_{in}}{dt}$ which could be used as the amplifier in a process control loop such as is shown in Figure 41.2. This computer or controller acts so as to correct any error or difference between, say, the set point temperature and the measured process temperature and also to add in extra corrective action proportional to the trend or derivative of the process output. This derivative action gives more rapid correction of process deviations from the set point because it can be considered to correct for the error which will occur if the present trend continues. A good analogy is that in hitting a ball in tennis, you aim at where the ball will be, not at where the ball is. You include a correction for the rate of change of position of the ball.

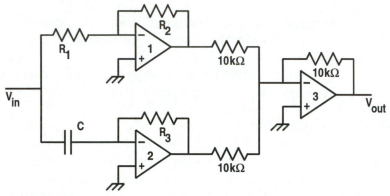

Figure 47.1: PD or proportional plus derivative action controller.

A suitable circuit for this application is shown in Figure 47.1. In this circuit, op-amp 1 gives the gain of $3.3V_{in}$ and so $\frac{R_2}{R_1} = 3.3$. If we arbitrarily choose $R_1 = 10\,\text{k}\Omega$ then $R_2 = 33\,\text{k}\Omega$.

Op-amp 2 gives the derivative action, $2.2\frac{dV_{in}}{dt}$, so we have $R_3C = 2.2$. We choose $C = 1\,\mu\text{F}$ and then get $R_3 = 2.2\,\text{M}\Omega$.

The two signals are then added in the inverting adder using op-amp 3. There are two stages of inversion, so the sign of the output signal is correct.

47.2   Problems

47.1 Design a circuit which has an output which is given by:

$$V_{out} = 23V_1 - 12V_2 + 8V_3$$

where V_1, V_2 and V_3 are inputs to the circuit. What is the minimum number of op-amps required?

47.2 Derive the mathematical expression which gives the output of the circuit, shown in Figure 47.2, in terms of the single input V_{in}.

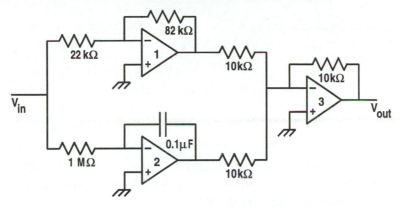

Figure 47.2: Problem 47.2.

47.3 A circuit has three inputs, V_1, V_2 and V_3. Design a circuit which will give an output which is described by:

$$V_{out} = 1.9V_1 - 2.8V_2 + 8.2V_3 + 0.6\frac{dV_1}{dt} - 3.6\int V_2 dt$$

47.4 The circuit shown in Figure 47.3 is used to move the scanning tip up and down as it scans the surface in the scanning tunnelling microscope manufactured by Burleigh Instruments. The input signal is the tip tunnelling current. The output signal drives the tip in the z direction. Explain the function of each of the circuit blocks.

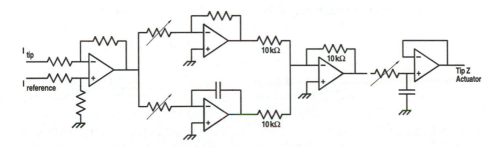

Figure 47.3: STM scanning tip height servo control circuit.

Unit 48   Active filters

- Active filters contain active devices such as transistors or op-amps so as to give gain as well as filtering action.

- The main advantage of active filters is that their performance can be made to be more independent of the signal source and load impedances.

- An iterative process is usually used to choose the best of the many possible designs for a particular application.

The passive filters examined in earlier units have the advantage of simplicity but their performance depends on the output impedance of the signal source and the input impedance of the load. This means that the corner frequencies can be changed when the passive filter is connected to different circuits.

More reliable performance is achieved by using active filters; that is, filters which include amplifiers such as op-amps to give some gain as well as filter action.

There are many hundreds of active filter designs which are well documented in specialist texts. We will examine one representative design which uses op-amps and which is suitable for use in the audio frequency region of the spectrum, that is for frequencies below about 10 kHz. These particular designs have been chosen for discussion because the author has used them in many applications and always found them to give consistent performance.

The three filters examined are low pass, high pass and band pass filters. The circuits for each type are given in Figures 48.1, 48.2 and 48.3. A useful feature of the design is that the shape of the circuit is the same in each case and that it is possible to use the same printed circuit board with different components to construct all three filters. This means that pcbs can be held ready in stock and used for different purposes, as needed.

A set of equations is given for each filter. In an RC filter there are many possible values for R and C which give a particular value of the RC product. Good design and also some experience will lead the user to avoid extreme values of R and C. In designing these filters you should try to use values of R within the range $100\,\Omega$ to $1\,\mathrm{M}\Omega$ and values for C within the range $1\,\mathrm{nF}$ to $1\,\mu\mathrm{F}$.

The design procedure is to make a guess at a reasonable value for some of the components, as indicated in each case, and then follow through the calculation to see if reasonable values of the other components are obtained from the calculations. The calculation is then repeated, using improved starting values until a satisfactory design is obtained. This iteration is best carried out by using a computer program or a programmable calculator. The writing of a suitable program is left as an exercise for the reader.

Besides the quantities, frequency, bandwidth and gain, which we have already met, we have the new parameter called peaking factor, α. The peaking factor describes the sharpness of the edge of a low or high pass filter. Some typical values for the increase in the gain at the band edge for high pass and low pass filters are shown in the table below.

α	0.1	0.3	1.0	1.4
dB peaking	20	10	3	0
Gain increase	10	3.16	1.4	1.0

Low pass filter

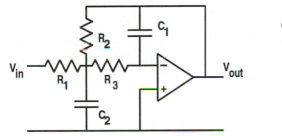 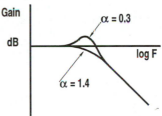

Figure 48.1: Low pass filter.

$$
\begin{aligned}
\text{Specified design quantities} \quad f &= \text{Cutoff frequency, Hz} \\
\alpha &= \text{Peaking factor} \\
A &= \text{Gain} \\
\text{Initial estimate} \quad C_1 \text{ in } \mu\text{F} \\
\text{Then} \quad C_2 &= \frac{4(1+A)C_1}{\alpha^2} \,\mu\text{F} \\
R_2 &= \frac{\alpha \times 10^6}{4\pi f C_1} \\
R_1 &= \frac{R_2}{A} \\
R_3 &= \frac{R_2}{1+A}
\end{aligned}
$$

High pass filter

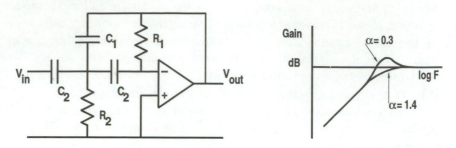

Figure 48.2: High pass filter.

Specified design quantities $\quad f \;=\;$ Cutoff frequency, Hz

$\qquad\qquad\qquad\qquad\qquad\alpha \;=\;$ Peaking factor

$\qquad\qquad\qquad\qquad\quad\; A \;=\;$ Gain

Initial estimate $\quad C_2 \;\;$ in $\;\mu F$

Then $\quad C_1 \;=\; \dfrac{C_2}{A}\,\mu F$

$$R_1 \;=\; \frac{(2A+1)\times 10^6}{2\pi f \alpha C_2}$$

$$R_2 \;=\; \frac{\alpha A \times 10^6}{2\pi f C_2 (2A+1)}$$

Band bass filter

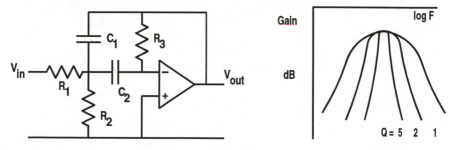

Figure 48.3: Band pass filter.

Specified design quantities $\quad f \;=\;$ Centre frequency, Hz

$\qquad\qquad\qquad\qquad\qquad B \;=\;$ Bandwidth at 3dB down from peak

$\qquad\qquad\qquad\qquad\qquad A \;=\;$ Gain

$\qquad\qquad$ Initial estimate $\quad C_1 \;$ in $\; \mu F$

$\qquad\qquad\qquad\qquad\qquad C_2 \;$ in $\; \mu F$

$$\text{Then} \quad Q \;=\; \frac{f}{B}$$

$$R_1 \;=\; \frac{Q \times 10^6}{2\pi f A C_1}$$

$$R_2 \;=\; \frac{1}{2\pi f Q (C_1 + C_2) \times 10^{-6} - \frac{1}{R_1}}$$

$$R_3 \;=\; \frac{Q \times 10^6}{2\pi f} \left(\frac{1}{C_1} + \frac{1}{C_2} \right)$$

48.1   Example

48.1 Design a low pass filter for a cutoff frequency of 1500 Hz, a gain of 12 and 3 dB of peaking at the band edge.

Use the circuit of Figure 48.1. A peaking of 3 dB at the band edge corresponds to a peaking factor of $\alpha = 1$, from the table. Make an initial estimate for C_1 and use the formulae for C_2, R_1, R_2 and R_3. The calculations can be presented as a table of successive iterations.

Iteration	1	2	3	4	5	6	Units
C_1	1.0	0.1	0.01	0.001	0.0005	0.0001	μF
C_2	52	5.2	0.52	0.052	0.026	0.005	μF
R_1	4.4	44	442	4.4k	8.8k	44k	Ω
R_2	53	530	5.3k	53k	106k	530 k	Ω
R_3	4.1	41	410	4.1k	8.2k	41k	Ω

In principle, all of these calculations give valid and correct filter design values.

In practice, iterations 1, 2 and 3 have such low resistances that the filter input impedances will be too low, typical op-amps will not be able to drive the large currents through the low resistances and the power consumption will be unnecessarily high.

Iteration 6 and any further iterations with smaller values of C_1 have overlarge values of R and the assumptions in the two rules for op-amps in Unit 39 begin to fail to be satisfied. Also high value resistors are unstable owing to current leakage in the thin moisture layer on the external surface of the resistors. If the capacitance of C_1 is too small, stray capacitances may come to dominate the operation of the circuit.

Therefore we reject otherwise valid designs when the resistances and capacitances are too small or too large. The circuit designer has to exercise judgment based on experience which is only sometimes codified in formal design rules. In this example, iteration 4 is a reasonable compromise.

The filter is fed with a signal from a signal source which itself has an output resistance. In principle, this source output resistance should be included in the value of R_1, otherwise the filter performance may be changed. However, we do not know the signal source resistance, so it is good practice to isolate the filter from the signal source by using a voltage follower stage in the input which has a high input impedance and low output impedance.

A suitable circuit, which includes this voltage follower input stage and which meets the stated requirements for frequency, peaking factor and gain, is shown in Figure 48.4. It should be noted that the calculated values for the resistors have been replaced by the nearest standard preferred value resistors.

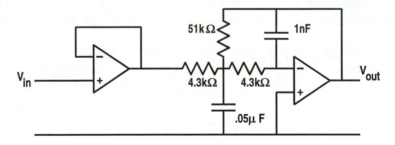

Figure 48.4: Example 48.1.

48.2   Problems

48.1 Design a low pass filter which has a cutoff frequency of 450 Hz, no peaking (0 dB) and a gain of 25. Draw your circuit design, showing the component values.

(These design values are given for reference purposes, to enable you to check your calculation method or program. They are not necessarily the best designs. $C_1 = 0.01\,\mu\text{F}$, $C_2 = 0.5\,\mu\text{F}$, $R_1 = 1\,\text{k}\Omega$, $R_2 = 25\,\text{k}\Omega$, $R_3 = 1\,\text{k}\Omega$.)

48.2 Design a high pass filter which has a cutoff frequency of 80 Hz, a peaking of 10 dB and a gain of 6. Draw your circuit design, showing the component values.

(Reference design: $C_1 = 0.033\,\mu\text{F}$, $C_2 = 0.2\,\mu\text{F}$, $R_1 = 430\,\text{k}\Omega$, $R_2 = 1377\,\Omega$.)

48.3 Design a band pass filter which has a centre frequency of 1.8 kHz, a bandwidth of 500 Hz and a gain of 15. Draw your circuit design, showing the component values.

(Reference design: $C_1 = 0.01\mu\text{F}$, $C_2 = 0.01\mu\text{F}$, $R_1 = 2122\,\Omega$, $R_2 = 2914\,\Omega$, $R_3 = 63.6\,\text{k}\Omega$.)

48.4 Is it possible to design a band pass filter, using this design for a band pass filter, which has a centre frequency of 300 Hz, a gain of 17 and a bandwidth of 10 Hz?

48.5 A particular microphone has a frequency response which is flat to within 3 dB from 16 Hz to 16 kHz with a 20 dB per decade drop off outside this range. The microphone is connected to the input of a low pass filter having a corner frequency at 13 kHz and an $\alpha = 1$. Calculate and sketch the frequency response of the system.

48.6 Write a computer program which prompts the user to select the type of filter, the required filter parameters and initial estimates of C_1 and C_2. The program should then compute the resistor values and print the list of component values to the screen.

Unit 49 Frequency response of op-amps

- The open loop frequency response of a 741 op-amp has a corner at 10 Hz and 100 dB.

- The open loop gain decreases by 20 dB for every factor of 10 increase in frequency above 10 Hz.

- The frequency response of a real amplifier is determined by the lower of:

  - The theoretical response set by the feedback components.
  - The response of the op-amp.

We have, until now, assumed that op-amps amplify signals at all frequencies with equal efficiency. This is not the case.

Since an op-amp has a high open loop gain, any spurious positive feedback from the output to the noninverting input can cause the amplifier to oscillate violently from full positive output to full negative output. Such positive feedback occurs most easily at high frequencies because the impedance of any stray capacitance between the output and the input causing positive feedback decreases with increasing frequency. Op-amps are therefore designed to have an open loop gain which decreases with increasing frequency so as to reduce the risk of positive feedback at higher frequencies. This is achieved by including a capacitor, fabricated on the silicon chip, in the op-amp internal circuit. This capacitance can be seen in the 741 op-amp circuit shown in Figure 35.5.

The open loop voltage gain of a typical op-amp such as the 741 is shown in Figure 49.1. This characteristic gain curve is very important and you should be able to sketch it without having to refer to Figure 49.1. Essentially the open loop gain is constant at 100 dB from DC up to 10 Hz. From 10 Hz upwards in frequency, the gain decreases by 20 dB for every factor of 10

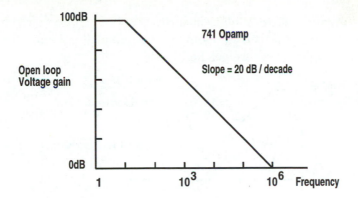

Figure 49.1: Open loop gain as a function of frequency for the 741.

increase in frequency, reaching a gain of $0\,\mathrm{dB}$ at a frequency of $10^6\,\mathrm{Hz}$. The key feature is the corner at $10\,\mathrm{Hz}$ and $100\,\mathrm{dB}$.

This curve is measured using a circuit such as that in Figure 49.2 in which the signal is applied directly between the two inputs to the op-amp without any negative feedback being used; hence the term open loop.

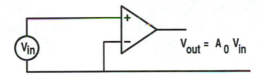

Figure 49.2: Circuit used to measure open loop gain.

Now let us examine how this op-amp frequency response affects the response of amplifiers built using op-amps.

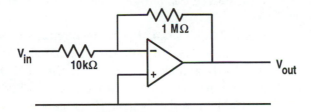

Figure 49.3: Inverting amplifier.

Figure 49.3 shows an inverting amplifier which has a gain of:

$$A = 20\log\left(\frac{1\,\mathrm{M\Omega}}{10\,\mathrm{k\Omega}}\right) = 40\,\mathrm{dB}$$

In theory the gain is 40 dB at all frequencies. However, if we take the op-amp response and draw a line corresponding to a nominal 40 dB amplifier gain on the op-amp response curve we get Figure 49.4.

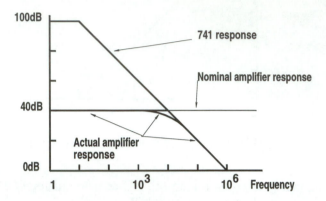

Figure 49.4: Concatenation of 741 response with amplifier response.

For frequencies below 10^4 Hz, the nominal amplifier response is below the op-amp open loop response so there is gain to spare for negative feedback. For frequencies above 10^4 Hz, the nominal gain is greater than the op-amp gain so the amplifier response is limited by the op-amp response and follows the op-amp response downwards. This is indicated as the actual amplifier response on the curve.

The general procedure is therefore to plot the gain as determined by the feedback components onto the op-amp response and then to obtain the actual response as the envelope of the lower of the two curves. The shape of the response is similar to that of an *RC* low pass filter not only in the dB response shape but also in the fact that on the downward part of the response there is a phase shift of 90° between the input and output signal at these high frequencies.

The bandwidth is usually defined as the frequency range over which the response is constant to within 3 dB. In this example the bandwidth extends from DC or 0 Hz up to 10 kHz and the bandwidth is then 10 kHz.

It can easily be seen that if resistors are chosen to give a high gain amplifier the penalty is that the bandwidth is low. If the gain is low then a high bandwidth is available.

The op-amp frequency response also affects the performance of the active filters examined in Unit 48. With active filters it is best to use a moderate gain so that the frequency response is determined by the filter design and not by the op-amp response. It is easy to obtain broadband gain with a later amplifier stage. It can also be seen that op-amps such as the 741 are not of significant use at frequencies above about 100 kHz. Special high frequency op-amps are available for high frequency work but are more expensive and are much more prone to instability unless great care is taken with the circuit design and layout.

49.1 Problems

49.1 Calculate the gain of the amplifier in Figure 49.5. Sketch the frequency response for the amplifier. Determine the bandwidth of the amplifier.

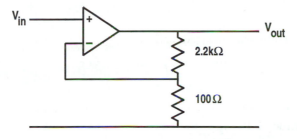

Figure 49.5: Problem 49.1.

49.2 What is the bandwidth of an inverting amplifier which uses a 741 op-amp and has a gain of -50? Sketch a suitable circuit with component values.

49.3 Design a noninverting amplifier which has a bandwidth from DC to 300 Hz. What will be the maximum gain that can be obtained if a 741 op-amp is used?

49.4 Derive an expression for the magnitude of the ratio of the output to input signals for sinusoidal signals for the differentiator circuit shown in Figure 49.6. Plot this response on the frequency response curve for the 741 op-amp and obtain the amplifier frequency response.

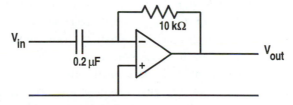

Figure 49.6: Problem 49.4.

Unit 50 Noise

- Noise signals present within a bandwidth B are specified in units of:

$$\text{Volts per } \sqrt{\text{Hz}} \quad \text{or} \quad \text{Amps per } \sqrt{\text{Hz}}$$

- The thermal or white noise from a resistor, R, at temperature T within a bandwidth B is:

$$V_{noise} = \sqrt{4kTRB}$$

- The shot noise associated with a DC current I is:

$$I_{noise} = \sqrt{2eIB}$$

- Flicker noise (sometimes called 'one over f noise') is comparable in magnitude to thermal noise at about $100\,\text{Hz}$ and has a spectrum which varies as $\frac{1}{f} = \frac{1}{\text{Frequency}}$.

If the gain of an electronic amplifier, with no signal present at the input, is gradually increased, a point is reached when the zero signal at the output starts to fluctuate or become noisy. This effect can be easily noted on audio amplifier systems when the noise appears as a hiss from the loudspeaker when the volume is turned up fully with no tape or CD inserted. The phenomenon can also be observed as snow on the screen of older TV sets when there is no signal at the input or when no antenna is plugged in. (More modern sets give the illusion of being noise free because of internal muting circuits which turn off the screen display when there is no signal at the input.)

Some of the origins of these noise signals are shown in Figure 50.1. On the left, we have man-made signals, loosely called interference. Good design and shielding of sensitive equipment in grounded metal cabinets or Faraday cages can prevent signals from reaching sensitive equipment. Legislation also requires manufacturers to reduce the spurious electromagnetic interference (EMI) emitted by equipment such as computers and household and office equipment. In the case of very sensitive equipment, moving the equipment to a quieter location can help to reduce the noise. However, there is an inherent limit to how quiet or noise free a system can be made. There

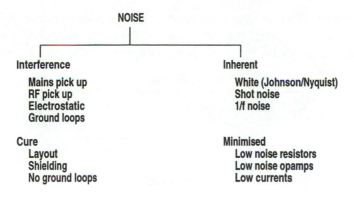

Figure 50.1: Noise sources.

are three unavoidable noise generation mechanisms present in all electronic systems as shown on the right in Figure 50.1.

Before examining the mechanisms in detail we must specify the units in which noise in electronic systems is measured.

By analogy with optical systems, we have a noise power spectrum. Light which has a power or intensity distribution which is uniform over the spectrum is called white light. Light which is more intense at longer (red) wavelengths or lower frequencies is said to be pink. We therefore have the concept of white noise having a power distribution spread equally over all frequencies.

We also have the concept of spectral density. If the power in a frequency range $\delta f = f_1 - f_2$ is measured, we can then plot the power per unit frequency as a function of frequency.

Since we are working with electronic systems in which we more conveniently measure the voltages and in which the power is proportional to the square of the voltage we use:

$$P = \frac{V^2}{R}$$

Then by measuring the noise voltages with an oscilloscope or voltmeter we have a quantity which is proportional to $\sqrt{P}$.

Consider the spectral density of this noise. At any given frequency, the units of power spectral density are watts per Hz bandwidth.

Therefore if we take the square root of the power spectral density we get the units of the noise voltage spectral density as:

Volts per $\sqrt{\text{Hz}}$

White noise was theoretically analyzed by Nyquist and experimentally measured by Johnson. Consider a resistor which is at a temperature T K. The electrons within the resistor move randomly with a kinetic energy appropriate to kT where k is Boltzmann's constant. Connect a resistor at each end of a matching transmission line as shown in Figure 50.2.

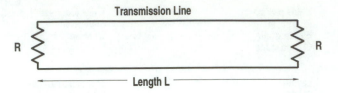

Figure 50.2: Transmission line carrying noise.

A good example of a transmission line is a coaxial cable used to carry radio or TV signals. Such a line has a characteristic impedance of 75 Ω which is determined by the diameters of the outer shield and the central conductor. So we will consider a coaxial cable terminated at each end by a 75 Ω matching resistor and we will also assume that the coaxial cable acts as a perfect loss free transmission line which does not itself introduce any noise.

The thermal motions of the electrons in the resistor cause the resistor to radiate electromagnetic signals into the transmission line and the resistor absorbs any signal travelling down the line to the resistor. After a short time, thermal equilibrium is reached and the spectral power distribution on the line is the same as the spectral distribution emitted by the resistor. Now cut the resistors from the line and leave the electromagnetic signal trapped on the line, bouncing back and forth between the ends.

What is the energy distribution of this signal on the transmission line? Or to put the question differently, what are the modes or frequencies that can be present as propagating waves on the line? The question is very similar to asking what modes of vibration can be present on a violin string or in an organ pipe.

If the length of the line is L and the velocity of the waves is C then the modes will be integer multiples of the fundamental, $f_0 = \frac{C}{2L}$.

The number of modes, N, between the frequency $f_1 = p\frac{C}{2L}$ and the frequency $f_2 = q\frac{C}{2L}$, where p and q are integers, is:

$$N = p - q = \frac{f_1}{\frac{C}{2L}} - \frac{f_2}{\frac{C}{2L}} = (f_1 - f_2)\frac{2L}{C}$$

Each mode has an energy kT associated with it and therefore the energy, W, on the line in the frequency range f_1 to f_2 is:

$$W = NkT = (p - q)kT = (f_1 - f_2)\frac{2L}{C}kT$$

This energy was delivered into the line by the two resistors in one line transit time $\frac{L}{C}$ and therefore the power from each resistor is:

$$P = \frac{W}{\frac{2L}{C}} = \frac{(f_1 - f_2)\frac{2L}{C}}{\frac{2L}{C}}kT = (f_1 - f_2)kT = \overline{i^2}R$$

where $\overline{i^2}$ is the mean square noise current.

The mean square noise voltage which drives this current through the two resistors in series is then (by squaring Ohm's law and making a substitution for $\overline{i^2}R$):

$$\overline{v^2} = \overline{i^2}(2R)^2 = 4\overline{i^2}RR = 4(f_1 - f_2)kTR = 4kTRB$$

where $B = f_1 - f_2$ is the bandwidth.

The noise voltage is then:

$$v_{noise} = \sqrt{4kTRB}$$

Shot noise. An electric current is numerically equal to the number of electrons flowing past a given point per unit time multiplied by the electronic charge. If N electrons on average flow in a given time interval then the statistical fluctuation in this number is $\sqrt{N}$ so that a specific measurement will be in the range $N \pm \sqrt{N}$. If the bandwidth of the measuring amplifier is B then the sample time is given by $\Delta t = \frac{1}{B}$.

The average DC current is $\quad I = \dfrac{Ne}{\Delta t} = NeB$

The fluctuation in the current is $\quad \Delta I = eB\sqrt{N} = eB\sqrt{\dfrac{I}{eB}} = \sqrt{eIB}$

A more detailed argument based on statistical mechanics shows that there is an extra factor of 2 so that the shot noise is then:

$$\Delta I = \sqrt{2eIB}$$

Flicker noise or $\frac{1}{f}$ noise. As the name suggests, this noise mechanism is most significant at low frequencies when $\frac{1}{f}$ is large. This can be seen in Figure 50.3 where the noise voltages associated with flicker noise are masked by shot noise at frequencies above some corner frequency f_0. Typically this corner frequency is at about $100\,\text{Hz}$.

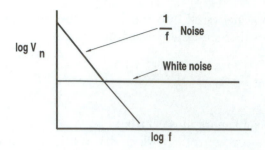

Figure 50.3: Noise spectra.

The precise mechanism by which flicker noise is generated is not fully understood but some progress has been made in modelling flicker noise by using the techniques of chaos theory and intermittency.

Flicker noise is a particular problem with DC amplifiers and very low frequency amplifiers when slow variations in signals on time scales of seconds are being examined. One technique for minimizing or avoiding flicker noise is to chop the signal at say 1 kHz and thus shift the the signal up in frequency and out of the flicker noise domain. Amplifiers which use this technique are called chopper amplifiers. The amplifiers used in medicine to measure the voltages associated with the heart (ECG) and brain (EEG) are examples of where flicker noise can be a problem.

50.1 Example

50.1 Calculate the thermal noise from the 500 kΩ resistor at 20°C which would be observed on an oscilloscope connected to the output of the circuit shown in Figure 50.4 (a). Assume that the op-amp does not introduce any noise. The V_S in the circuit diagram is the Thévenin equivalent of the noise source.

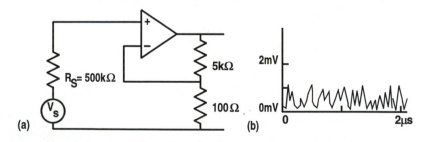

Figure 50.4: Example 50.1.

The gain of the amplifier is $A = 1 + \frac{5000}{100} = 51$ or 34 dB.

Plot the 34 dB line on the op-amp frequency response curve shown in Figure 49.1 and the bandwidth is obtained to be about 30 kHz.

The thermal noise from the 500 kΩ resistor is given by:

$$
\begin{aligned}
V_{noise} &= \sqrt{4kTR_s} \times \sqrt{\text{Bandwidth}} \\
&= \sqrt{4 \times 1.38 \times 10^{-23} \times 293 \times 5 \times 10^5 \times 3 \times 10^4} \\
&= 16\,\mu\text{V}_{\text{RMS}}
\end{aligned}
$$

This noise signal is amplified by a factor of 51 to give an output signal of 0.87 mV$_{\text{RMS}}$. On a good oscilloscope, this will appear as a fine 'grass' on the display when the scope is set at near maximum sensitivity. This is shown in Figure 50.4 (b)

50.2   Problems

50.1 Calculate the shot noise voltage within a 15 MHz bandwidth due to a current of 0.8 A flowing through a 50 Ω resistor. Could this noise be observed with an oscilloscope?

50.2 A radio receiver has a bandwidth of 10 kHz and a 300 Ω antenna input connection. The equivalent noise temperature at the antenna input due to the electronic circuits of the receiver is specified by the manufacturer to be 600 K. Calculate the voltage signal at the antenna terminals which will give a 10:1 signal to noise ratio.

50.3 A strain gauge is a metal foil resistor whose resistance changes when it is strained. Two such gauges are bonded onto opposite sides of a cantilever beam of thickness, T, which is loaded to a radius of curvature, D, as shown in Figure 50.5 (a). For such gauges $\frac{dR}{R} = G \times \frac{dL}{L}$ where G is the gauge factor and dL and dR are the changes in gauge length and resistance. The gauges used have unstrained resistance of 120 Ω and gauge factor, $G = 2.1$. Calculate the noise voltage which is present at the output of the bridge circuit shown in Figure 50.2 (b). Calculate the output voltage from the bridge for a given fractional change in the radius of curvature of the beam. Also calculate the fractional change in the radius of curvature which gives a signal to noise ratio of 1.

50.4 The noise generated in the internal circuits of a 741 op-amp can be described by two noise sources v_n and i_n connected to the input to an ideal noiseless op-amp as shown in Figure 50.6 (a). These sources

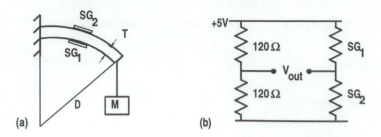

(a)

(b)

Figure 50.5: Problem 50.3. Strain gauge bridge.

have values of 4×10^{-15} V_{RMS}^2 Hz^{-1} and 5×10^{-25} A_{RMS}^2 Hz^{-1}. Use the principle of superposition and the two rules for op-amps in Unit 39 to calculate the noise signal at the output of the amplifier shown in Figure 50.6 (b). Note that the resultant, V_R of two random noise sources, V_1 and V_2 is given by $V_R = \sqrt{V_1^2 + V_2^2}$.

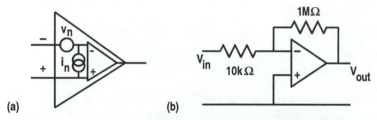

(a)

(b)

Figure 50.6: Problem 50.4. Op-amp noise evaluation.

50.5 Analyze the operation of the active full wave rectifier shown in Figure 50.7 and calculate the average output signal for a sinusoidal signal of 100 mV amplitude at 3 kHz and for a white noise signal, bandwidth limited to 3 kHz and 0.4 V_{RMS}. Sketch the output waveforms.

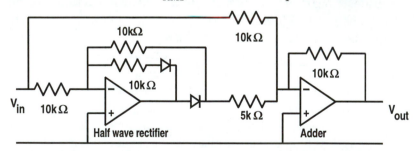

Figure 50.7: Problem 50.5. Precision full wave rectifier.

Unit 51   Recovery of small signals from noise

- **Instrumentation amplifiers** use a three op-amp configuration to obtain high input resistance and improved common mode rejection. The gain is set by a single gain setting resistance which can be chosen for high thermal stability.

- **Isolation amplifiers** have no galvanic connection between input and output and can operate with common mode voltages up to thousands of volts. They are also used for protecting patients from danger of electrical shock in medical applications of electronics.

- **Phase sensitive detectors** use the fact that the frequency and phase of a signal are known and that a reference signal is available to gain-switch from $+1$ to -1 and obtain synchronous detection of a signal which may be buried in noise.

There are many situations in electronics where signals are present but are partially masked by noise. It is not enough to increase the amplification since the noise is also amplified with the signal and the signal to noise ratio remains unchanged. There are three techniques or types of circuits which are frequently used when the signal is very small. The general aim is to obtain as good a signal as possible to start with and then to use the fact that the signal is at a known frequency and phase in order to improve the detection of the signal.

Instrumentation amplifier. The ordinary op-amp has good common mode rejection; that is, it is insensitive to signals which are present at both the inverting and noninverting inputs. However, when the signal is a small difference voltage in the presence of a large common mode signal then an instrumentation amplifier should be used. An example of such a signal would be two voltages from a bridge circuit, $V_1 = 1.000000\,\text{V}$ and $V_2 = 1.000001\,\text{V}$. If we wish to measure the difference between V_1 and V_2, which in this case is $1\,\mu\text{V}$ in the presence of the $1\,\text{V}$ common mode signal, then an amplifier using a single op-amp would have to have the two inputs matched to one part in a million, which is not possible.

The traditional instrumentation amplifier uses three op-amps in the circuit configuration shown in Figure 51.1.

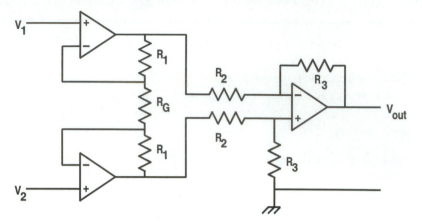

Figure 51.1: The three op-amp instrumentation amplifier.

Two op-amps are used symmetrically for the input amplifier followed by the differential amplifier which was analyzed in Unit 43. In the circuit in Figure 51.1, consider the resistor chain R_1, R_G and R_1. (Note the convention that resistors with the same labels have the same values.)

Using the rules for op-amps from Unit 39, the voltage across the gain setting resistor, R_G, is $V_1 - V_2$ and therefore the current in the resistor chain is:

$$I = \frac{V_1 - V_2}{R_G}$$

The voltage applied to the input to the differential amplifier is then:

$$V_{diff} = \frac{V_1 - V_2}{R_G} \times (R_1 + R_G + R_1)$$

The differential amplifier has a gain of $\frac{R_3}{R_2}$ and therefore the output from the instrumentation amplifier is:

$$V_{out} = \frac{2R_1 + R_G}{R_G} \times \frac{R_3}{R_2} \times (V_2 - V_1)$$

Typically an instrumentation amplifier having this configuration would be manufactured as a hybrid integrated circuit by specialist companies such as Analog Devices, using selected low noise, ultra stable op-amps. You would not normally construct an instrumentation amplifier from discrete components. The gain of the instrumentation amplifier is set by a single, external gain setting resistor having a high thermal stability. The instrumentation amplifier is usually supplied in a metal can rather than plastic package so as to have better shielding from interference. Such a device is capable of operating close to the inherent noise limits discussed in Unit 50.

Isolation amplifier. The ultimate in common mode rejection is achieved by an isolation amplifier which has zero common mode response because there is no galvanic connection between the input and output. These circuits are not normally constructed from components by the user but are also bought as complete devices from specialist manufacturers such as Analog Devices.

The most common configuration of isolation amplifier is a low power consumption instrumentation amplifier followed by a voltage to frequency converter in which the frequency output is proportional to the input differential voltage. This signal is then coupled through an isolating transformer which passes the high frequency signal but blocks any DC or low frequency AC. The secondary of the transformer is then fed to a frequency to voltage converter and to a low pass filter and the output. The basic configuration is shown in Figure 51.2.

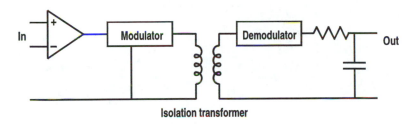

Figure 51.2: Isolation amplifier using a transformer for isolation.

There is usually a second transformer in the system which couples AC power to a rectifier which powers the input stage, again without having any galvanic connection with the input stage. This second transformer is not shown in the diagram.

One very important application for these isolation amplifiers is for providing isolation between the electrodes connected to patients and the mains powered equipment in hospitals. These electrodes are connected for electrocardiography, ECG (heart monitoring), and electroencephalography, EEG (brain wave monitoring). In the event of a fault in the mains powered equipment, the isolation amplifier prevents the patient receiving a shock. Also, because of the isolation, it is possible to use heart resuscitation, electro shock equipment without burning out the sensitive amplifiers.

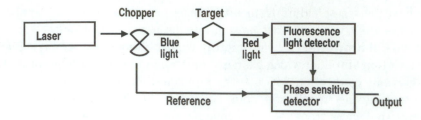

Figure 51.3: Chopper and phase sensitive detector configuration.

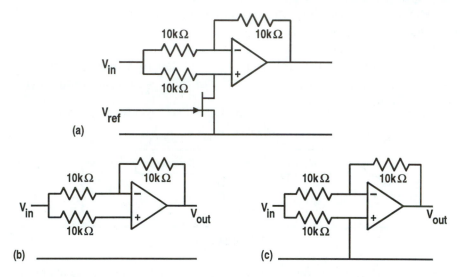

Figure 51.4: Basic phase sensitive detector circuit.

Phase sensitive detection. If the amplitude of an AC waveform is to be measured, a very simple technique is to use a diode rectifier and measure the DC voltage at the output of the rectifier (see Unit 28). However, if there is noise present as well as signal then the noise will also be rectified and contribute to the output. There is no way a simple rectifier can distinguish between signal and noise and a signal to noise ratio of at least 10:1 is necessary for valid detection of a signal using a diode detector.

There are many situations where a signal at a particular frequency is present but is partially or fully masked by noise and where the signal frequency is available for reference. An example of this situation is shown in Figure 51.3 where a laser beam is chopped at frequency, f_R, by a rotating chopper blade. The laser beam subsequently causes a weak fluorescence in a target material and the fluorescent light is then detected by a photodetector. In fluorescence, short wavelength light is absorbed in a target. This causes excitation of the molecules in the target. The excited molecules rapidly release some of this energy as lower energy, longer wavelength light. The signal from the photodetector may be noisy but it is known that if a valid fluorescence signal is present then it will be at a frequency f_R because this is the frequency at which the exciting laser light is modulated. This is the type of situation where a phase sensitive detector would be used. The phase sensitive detector essentially uses the information that the signal is at a particular frequency to pull the signal out of the noise background.

The basic electronic configuration of a phase sensitive detector is shown in Figure 51.4 (a).

A reference square wave signal from $0\,V$ to $-5\,V$ is available from the chopper which is synchronized with the signal to be detected. This signal is applied to the gate of the FET.

When the reference is at $-5\,V$ the FET is off or nonconducting and the effective circuit is that shown in Figure 51.4 (b). No current flows into the noninverting input of the op-amp and the noninverting input is at the signal voltage. The output is fed back to the inverting input and the amplifier therefore has a gain of $+1$.

When the reference voltage is at $0\,V$ the FET conducts and the effective circuit is that shown in Figure 51.4 (c). This is an inverting amplifier with a gain of -1. (The $10\,k\Omega$ resistor from the signal input to the noninverting input and ground does not affect the operation of the amplifier.)

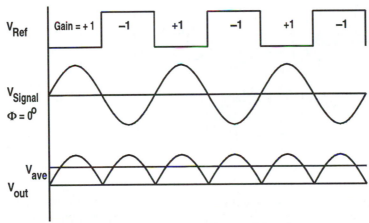

Figure 51.5: Reference and signal in phase.

The circuit for the phase sensitive detector is therefore an amplifier which is switched between a gain of $+1$ and -1 by the application of the reference square wave.

Now consider what happens when there is a sinusoidal input signal which is at the same frequency as the reference and which is in phase with the reference, that is $\phi = 0°$. This situation is shown in Figure 51.5.

The upper waveform is the reference from the chopper. The corresponding value of the gain of the amplifier is shown for each half cycle of the reference. The signal is then multiplied by this value of the gain to give the output in the third waveform. It can be seen that the negative half of the signal waveform is inverted to give a positive half cycle. The net effect is similar to having a diode rectifier with one important difference. Any noise will be uncorrelated with the reference waveform and therefore a pure noise input will give an average output of zero whereas a signal at the reference frequency and having zero phase difference will give an average output as shown on the diagram as V_{ave}. Typically this averaging is carried out using an RC low pass filter or an op-amp integrator.

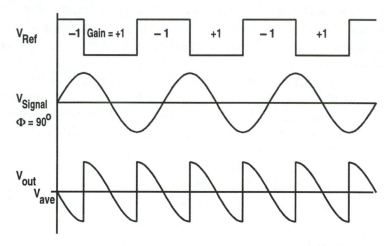

Figure 51.6: Reference and signal 90° out of phase.

The phase sensitive aspect of the circuit becomes apparent when we look at the situation when the phase difference between the reference and the signal is $\phi = +90°$ as is shown in Figure 51.6.

Again the signal is multiplied by either $+1$ or -1 depending on the phase of the reference waveform and the resultant waveform is shown at the bottom. In this case when the phase difference is 90° it can be seen that the output averages to zero. This is the phase sensitive aspect of the detection system.

The case when the phase difference is $\phi = 180°$ is shown in Figure 51.7. The output voltage now averages to a negative value.

So the output goes from full positive output to full negative output as the relative phase of the reference and the signal varies from 0° to 180°.

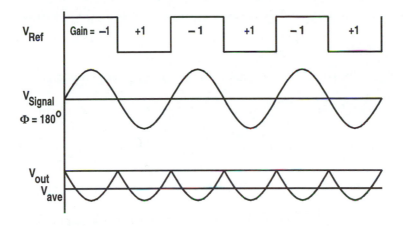

Figure 51.7: Signal and reference 180° out of phase.

In a commercial phase sensitive detection system there is usually a control on the front panel which allows you to vary the phase electronically over a range of 360° so as to obtain signal maxima when the phase is 0° and 180°. No maximum will be obtained for noise or for signals at frequencies different from the reference frequency. It is therefore possible to measure signals which are buried in noise. In other words, it is possible to operate with fractional signal to noise ratios. The commercial implementations of combined amplifiers and phase sensitive detectors are usually called lock-in amplifiers and these systems are capable of measuring voltage signals down to levels of nanovolts $(10^{-9}\,\text{V})$

51.1 Problems

51.1 What is the bandwidth of a low pass filter having an RC time constant of 2 seconds?

51.2 If the noise at the input to an amplifier is $740\,\mu\text{V}$ per $\sqrt{\text{Hz}}$, what would be the RMS noise voltage measurement if the signal were passed through a low pass filter having a time constant of 5 seconds?

51.3 The input signal to a lock-in amplifier consists of a signal of $23\,\mu\text{V}$ mixed with a noise signal of $120\,\mu\text{V}$ per $\sqrt{\text{Hz}}$. Calculate the signal to noise ratios for filter time constants of 0.1 seconds and 2 seconds.

Unit 52   Timers

- The voltage across a capacitor in a charging RC circuit is:

$$V_C = V_{sup}\left(1 - e^{\frac{-t}{RC}}\right)$$

- The voltage across the capacitor in a discharging RC circuit is:

$$V_C = V_{sup}e^{\frac{-t}{RC}}$$

- The charge-up time or discharge time between $\frac{1}{3}V_{sup}$ and $\frac{2}{3}V_{sup}$ is :

$$T = 0.7RC$$

- The time taken for a capacitor to charge from $0\,\text{V}$ to $\frac{2}{3}V_{sup}$ is :

$$T = 1.1RC$$

- The two time intervals generated by a 555 Timer IC are given by:

$$T_1 = 0.7(R_A + R_B)C \quad \text{and} \quad T_2 = 0.7R_BC$$

The basic timing element used in integrated circuit timers such as the 555 Timer is the RC charging circuit shown in Figure 52.1 (b).

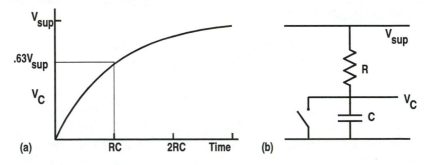

Figure 52.1: RC charging curve.

The switch shorts out the capacitor until time $t = 0$ when the switch is opened allowing the capacitor to charge up through the resistor. The voltage across the capacitor, as a function of time, is shown in Figure 52.1 (a).

The analysis proceeds as follows:

The voltage across the resistor R is $V_{sup} - V_C$.

The current is therefore $I = \frac{dQ}{dt} = \frac{V_{sup} - V_C}{R}$.

The charge on the capacitor is $Q = CV_C$ which, after differentiation, becomes $\frac{dQ}{dt} = C\frac{dV_C}{dt}$.

Equate these two expressions for the current to get:

$$RC\frac{dV_C}{dt} = V_{sup} - V_C$$

This differential equation has a solution:

$$V_C = V_{sup}\left(1 - e^{\frac{-t}{RC}}\right)$$

where RC is called the time constant. This is the graph of the voltage across the capacitor which is plotted in Figure 52.1 (a).

At the time $T = RC$, the calculated value of the V_C is $0.63V_{sup}$. This is indicated on the graph of Figure 52.1 (a). The output has made 63% of its total change at time $T = RC$.

This 63% response in one time constant has applications in other areas of measurement since many measuring instruments have a similar first order time response. A thermometer is warmed by heat transfer from the surroundings. The rate of heat transfer is proportional to the temperature difference between the thermometer and the surroundings. The more massive the thermometer (and usually the more rugged the thermometer) the larger the heat capacity or time constant. A typical industrial thermometer will have a time constant of about 30 seconds. This means that if the temperature of the flowing fluid in which the thermometer is immersed changes suddenly, it will take 30 seconds before the thermometer registers 63% of the change and 120 seconds before the reading is within 2% of the correct value.

If a slightly different circuit is used, such as that in Figure 52.2 (b), then when the switch is open the voltage across the capacitor is the supply voltage V_{sup}. If the switch is closed at time $t = 0$ then the capacitor discharges through R following the curve shown in Figure 52.2 (a). A differential equation can be set up as in the analysis of the RC charging case and the equation:

$$V_C = V_{sup}e^{\frac{-t}{RC}}$$

obtained for the voltage across the capacitor.

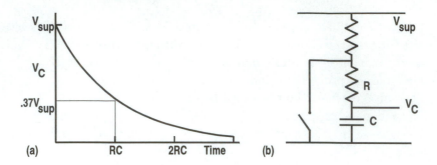

Figure 52.2: *RC* discharging curve.

Define T_L as the time taken for the capacitor in Figure 52.3 (b) to charge from $0\,\mathrm{V}$ to the lower voltage V_L.

Define T_H as the time taken for the capacitor in Figure 52.3 (b) to charge from $0\,\mathrm{V}$ to the higher voltage V_H.

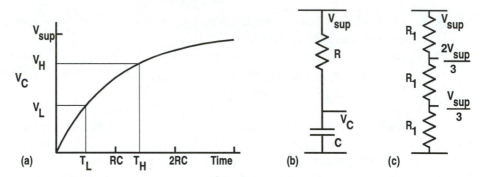

Figure 52.3: Charge-up time to the two voltages V_L and V_H.

From the solution to the charge-up equation we get:

$$V_L = V_{sup}\left(1 - e^{\frac{-T_L}{RC}}\right)$$

$$\text{and} \quad V_H = V_{sup}\left(1 - e^{\frac{-T_H}{RC}}\right)$$

Rearrange these equations and take natural logs (ln) to get:

$$-T_L = RC\ln\left(\frac{V_{sup} - V_L}{V_{sup}}\right)$$

$$\text{and} \quad -T_H = RC\ln\left(\frac{V_{sup} - V_H}{V_{sup}}\right)$$

Subtract these two to get:

$$T_H - T_L = RC \left(\ln \left(\frac{V_{sup} - V_L}{V_{sup}} \right) - \ln \left(\frac{V_{sup} - V_H}{V_{sup}} \right) \right) = RC \ln \left(\frac{V_{sup} - V_L}{V_{sup} - V_H} \right)$$

Now set $V_L = \frac{1}{3}V_{sup}$ and $V_H = \frac{2}{3}V_{sup}$.

We have chosen these values because it is easy to get these two voltages by using three resistors in series connected between the power supply and ground as shown in Figure 52.3 (c).

Put these voltages into the expression for $T_H - T_L$ to get a time T such that:

$$T = RC \ln \left(\frac{V_{sup} - \frac{1}{3}V_{sup}}{V_{sup} - \frac{2}{3}V_{sup}} \right) = RC \ln 2 = 0.693 RC \approx 0.7 RC$$

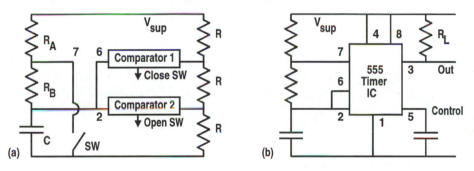

Figure 52.4: Internal circuit blocks of the 555 Timer.

Consider the circuit shown in block diagram form in Figure 52.4 (a). Three resistors in series give reference voltages of $\frac{1}{3}V_{sup}$ and $\frac{2}{3}V_{sup}$. The voltage across the capacitor is compared with these two voltages in comparators 1 and 2. The outputs of the two comparators toggle a switch SW which changes the capacitor from being charged up through R_A and R_B in series, with time constant $(R_A + R_B)C$, to being discharged through R_B, with time constant $R_B C$. When the voltage across the capacitor increases past $\frac{2}{3}V_{sup}$ comparator 1 toggles and closes the switch, starting a discharge cycle. When the voltage across the capacitor decreases below $\frac{1}{3}V_{sup}$, comparator 2 toggles and opens the switch to allow the capacitor to charge up again. The capacitor is therefore alternately charged up and discharged in a continuous cycle between the limits of $\frac{1}{3}V_{sup}$ and $\frac{2}{3}V_{sup}$.

In the diagram of Figure 52.4 (a), the numbers at the various points in the circuit indicate the pin numbers of the 555 Timer IC which is shown in Figure 52.4 (b). The comparators, the three resistor chain and the switch are all contained within the 555 Timer IC.

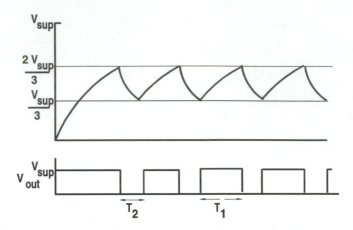

Figure 52.5: Capacitor and output voltage waveforms for the 555 Timer.

The voltage across the capacitor as a function of time is shown in Figure 52.5 which also shows the section of the charging and discharging exponential curve followed by the capacitor voltage. The 555 Timer has one extra important feature. Pin 3 gives an output signal which is an indicator of the state of the IC comparators. The output from pin 3 is at the supply voltage when the capacitor is charging up and the output from pin 3 is at $0\,$V when the capacitor is discharging.

The timing of the waveform is such that the time T_1 for the capacitor to charge from $\frac{1}{3}V_{sup}$ to $\frac{2}{3}V_{sup}$ is given by:

$$T_1 = 0.7(R_A + R_B)C$$

because the capacitor charges through R_A and R_B in series. The time for the capacitor to discharge from $\frac{2}{3}V_{sup}$ to $\frac{1}{3}V_{sup}$ is given by:

$$T_2 = 0.7R_BC$$

because the capacitor discharges through R_B.

One disadvantage of this timer circuit is that the output from pin 3 is an asymmetric waveform. The mark to space ratio is not 1:1, that is the waveform is not square. Greater control of the mark to space ratio can be achieved by using a circuit such as that shown in Figure 52.6.

Here the capacitor charges up through R_1 and the diode with a time constant of $T_1 = 0.7R_1C$.

The discharge path is through R_2 and the diode in series with R_2 to give a time constant $T_2 = 0.7R_2C$.

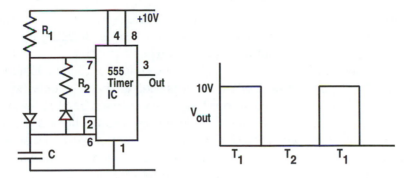

Figure 52.6: Circuit for generation of arbitrary mark to space ratio.

The time constants are now independently set because the steering diodes steer the charging and discharging currents through different resistors. Any mark to space ratio can be obtained with this circuit.

These applications of the 555 Timers have all been as astable oscillators where a continuous train of pulses is generated. In contrast, Figure 52.7 shows a circuit which gives monostable operation.

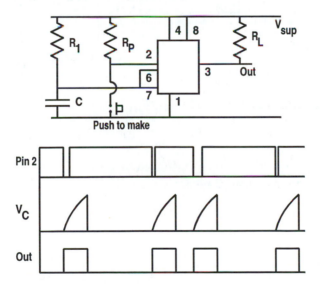

Figure 52.7: Triggered operation of the 555 Timer.

When the push to make button is pressed, the voltage on pin 2 is brought to 0 V and this triggers comparator 2 (refer to Figure 52.4 (a)) and causes the switch to open, allowing the capacitor to charge up. As soon as the voltage on the capacitor reaches $\frac{2}{3}V_{sup}$, comparator 1 operates and the switch is closed again, discharging the capacitor. The output voltage at pin 3 is normally at $0\,\mathrm{V}$ but rises to V_{sup} during the time the capacitor is charging up and a pulse of length $T = 1.1RC$ appears at the output each time the button is pressed. The length of this output pulse is independent of the length of time for which the button is pressed as long as it is shorter than the pulse length.

This circuit is a very convenient method of generating a fixed pulse length on demand. A typical application of this would be a controller for a photographic enlarger timer switch. Each time the button is pressed, the enlarger lamp turns on for a fixed length of time. The ON time can be adjusted by varying the time constant by using a variable resistor in place of the fixed resistor R_1.

52.1   Example

52.1 Calculate component values for a 555 Timer circuit which generates a $0\,\mathrm{V}$ to $10\,\mathrm{V}$ waveform with a $0\,\mathrm{V}$ for $6\,\mathrm{ms}$ and a $10\,\mathrm{V}$ for $13\,\mathrm{ms}$.

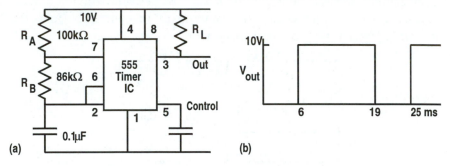

Figure 52.8: Example 52.1.

A suitable circuit is shown in Figure 52.8 (a). Choose a value for C of $0.1\,\mu\mathrm{F}$. The discharge time is then $6 \times 10^{-3}\,\mathrm{s}$ which gives:

$$T_2 = 6 \times 10^{-3} = 0.7 \times 0.1 \times 10^{-6} \times R_B$$

and therefore $R_B = \frac{6 \times 10^{-3}}{0.7 \times 0.1 \times 10^{-6}} = 86\,\mathrm{k\Omega}$.
We also have:

$$T_1 = 13 \times 10^{-3} = 0.7 \times 0.1 \times 10^{-6}(86\,\mathrm{k\Omega} + R_A)$$

and therefore $86\,\mathrm{k\Omega} + R_A = 186\,\mathrm{k\Omega}$ which gives $R_A = 100\,\mathrm{k\Omega}$.
The resulting waveform output from the circuit is shown in Figure 52.8 (b). The values of $100\,\mathrm{k\Omega}$ and $86\,\mathrm{k\Omega}$ for R_A and R_B depend on the initial choice of C. If R_A or R_B had turned out to be substantially different from these values then a different value for C might have to be selected.

52.2 Problems

52.1 Show that the time taken for a capacitor to charge from $0\,\text{V}$ to $\frac{2}{3}V_{sup}$ through a resistor R is given by $1.1RC$.

52.2 Design a circuit using a 555 Timer IC which gives an output waveform of $12\,\text{V}$ for $50\,\text{ms}$ followed by $0\,\text{V}$ for $16\,\text{ms}$. Sketch the expected waveform and the circuit showing component values.

52.3 Design a circuit using a 555 Timer IC which gives an output waveform of $9\,\text{V}$ for $0.1\,\text{ms}$ followed by $0\,\text{V}$ for $8\,\text{ms}$. Sketch the expected waveform and the circuit showing component values.

52.4 Design a circuit which will give a single pulse of $3.2\,\text{s}$ duration each time a button is pressed. Sketch the expected waveform and the circuit showing component values.

52.5 Design a circuit and calculate component values for a 555 Timer circuit which will flash a set of three high intensity light emitting diodes (leds) on and off once per second so as to operate as a compact bicycle rear light. The leds can be connected in parallel between pin 3 and the supply voltage. Discuss the relative advantages and disadvantages of putting the leds in series or in parallel.

52.6 Calculate the output, at pin 3, from the circuit shown in Figure 52.9. Sketch the voltage waveforms which would be observed with an oscilloscope when Channel A is connected to pin 3 and Channel B is connected across the capacitor.

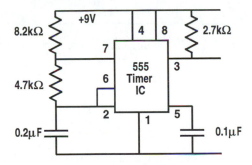

Figure 52.9: Problem 52.6.

Unit 53 Sinusoidal oscillators

- Sinusoidal voltage waveforms are obtained by using an amplifier

  - with positive feedback,
  - with a loop gain of 1 and
  - with a frequency selective feedback network.

Consider the amplifier shown in Figure 53.1. A sinusoidal signal which is applied to the input is phase shifted and attenuated in each of the three CR filters. Assume for simplicity that the three filters are noninteracting; that is, that we can calculate the effect of each filter without having to allow for loading effects of the other filters. Also select the transistor bias so that the input resistance of the transistor amplifier is equal to the value of the resistors used in the CR stages of the filters.

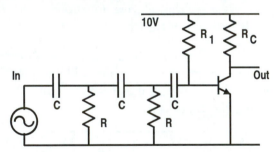

Figure 53.1: Phase shift network and amplifier.

The phase shift in a single CR stage is given by:

$$\phi_1 = \tan^{-1}\left(\frac{1}{2\pi fCR}\right)$$

The phase shift between the input and the base of the transistor is $3\phi_1$ and the value of ϕ_1 is between $0°$ and $90°$. The transistor amplifier inverts the signal as well as amplifying the signal. Inversion corresponds to a phase shift of $180°$. Therefore the total phase shift from the input to the amplifier to the output is $\phi_{total} = 3\phi_1 + 180°$. The output will then be in phase with the input but shifted by one period when $\phi_1 = 60°$, that is for a frequency $f_0 = \frac{1}{2\sqrt{3}\pi CR}$. (See Unit 13.)

Now remove the signal generator from the input and loop the output back to the input as shown in Figure 53.2.

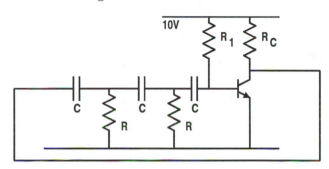

Figure 53.2: Phase shift oscillator.

If the loop gain is greater than 1, that is if the gain of the transistor amplifier more than compensates for the attenuation in the three CR filters, then any very small signal (noise) in the circuit at frequency $f_0 = \frac{1}{2\sqrt{3}\pi CR}$ will be amplified and phase shifted by 360° as it propagates around the loop with the result that the amplitude of a signal at frequency f_0 will grow with each loop traversal and the circuit will break into oscillation at frequency f_0. The amplitudes of the oscillations grow until they are limited by the power supply voltage. Since the only frequency satisfying the condition is $f_0 = \frac{1}{2\sqrt{3}\pi CR}$ we then have a sinusoidal function generator.

Here we have employed positive feedback to cause the circuit to go into oscillation. When the internal construction of a 741 op-amp was discussed in Units 35 and 49, it was pointed out that the capacitor included in the op-amp gave stability. The function of the internal capacitor in the 741 op-amp is to form a filter and prevent the phase shift ever reaching 180° and therefore prevent the 741 op-amp from going into unwanted oscillation due to positive feedback resulting from stray external capacitances coupling signal from the output back to the input.

If the circuit shown in Figure 53.2 is constructed and the output waveform is examined with an oscilloscope it will be found that the waveform is distorted from a sinusoidal waveform as is shown in Figure 53.3.

The reason for this distortion is that the transistor amplifier ceases to be fully linear when large signals are present. The solution to this nonlinearity distortion problem is to introduce a feedback mechanism which reduces the gain of the amplifier if the output amplitude becomes too large. This is called amplitude stabilization.

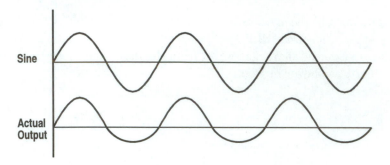

Figure 53.3: Distortion due to saturation nonlinearity.

In order to implement amplitude stabilization we need a device whose resistance changes smoothly as the current in the device increases. The simplest and cheapest such device is a small, low wattage filament bulb such as a T1-3 mm, 5 V, 50 mA bulb. When the current through the bulb increases, the filament heats up and the resistance of the filament also increases to $R_B = \frac{5\,\text{V}}{50\,\text{mA}} = 100\,\Omega$ when the bulb is at operating voltage. When the bulb is cold the resistance is typically about $9\,\Omega$. The thermal capacity of the filament is such that its resistance remains at the higher value from one half cycle to the next for frequencies above about 10 Hz.

Consider the operation of the potential divider circuit at the output of the circuit in Figure 53.4.

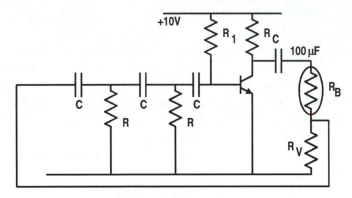

Figure 53.4: Amplitude stabilization by control of feedback.

The potential divider comprising the bulb, which has a resistance R_B, and R_V allows a fraction of the output sinusoidal signal given by $\frac{R_V}{R_B+R_V}$ to be fed back. R_V is adjusted so that the oscillator just oscillates with a clean sinusoidal waveform. If the oscillation amplitude grows then the voltage across the potential divider increases, the bulb heats owing to the increased current through the bulb and the resistance of the bulb R_B increases so that the positive feedback fraction given by $\frac{R_V}{R_B+R_V}$ decreases tending to reduce the amplitude of oscillation. The amplitude of the oscillation is thus regulated at one stable value which can be chosen to keep the amplifier operating in the linear region and keep a sinusoidal output waveform.

The same principles of positive feedback and amplitude stabilization can be employed in a Wein bridge oscillator which has a more readily adjustable frequency of operation and is therefore more suitable for use as a laboratory sinusoidal signal generator.

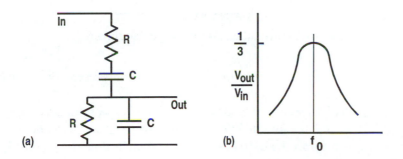

Figure 53.5: Wein bridge and response.

First consider a Wein bridge as shown in Figure 53.5 (a). This is a band pass filter which has a response curve as shown in Figure 53.5 (b). The frequency at the peak of the pass band is $f_0 = \frac{1}{2\pi RC}$.

In order to construct an oscillator we require that the loop gain be greater than 1. The attenuation at the peak of the Wein bridge response is $\frac{1}{3}$ and therefore the gain of the amplifier must be at least 3 for oscillation to take place or more than 3 when amplitude stabilization is employed.

The circuit in Figure 53.6 shows how the op-amp is set up with a Wein bridge. The output from the op-amp is fed back through the bridge with the output of the bridge going to the noninverting positive feedback input to the op-amp in order to give oscillation. The resistor, R_1, and the bulb resistor initially give very small amounts of negative feedback which allows the oscillator to start oscillating. The bulb resistance at this time is low because there is no current through the bulb at the start.

As the oscillation grows, the bulb heats up and the fraction of the output that is fed back to the inverting input or negative feedback input to the op-amp increases until an amplitude of oscillation is reached when the bulb is at a temperature which gives a loop gain of 1 and a clean sinusoidal signal is obtained at the output. The loop gain is less than 1 at other frequencies

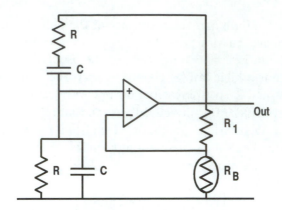

Figure 53.6: Wein bridge oscillator.

and therefore other Fourier components will not be present in the waveform thus giving a pure sinusoidal waveform.

Continuous frequency control by a factor of 10 is obtained by using a twin ganged potentiometer for the two Wein bridge resistors. Decade changes of frequency are obtained by switching in pairs of capacitors with a multiple pole switch. A potentiometer on the output allows the output voltage amplitude from the unit to be varied from $0\,\mathrm{V}$ to about $10\,\mathrm{V}$.

53.1 Examples

53.1 Design a transistor phase shift oscillator for a fixed frequency of 300 Hz.

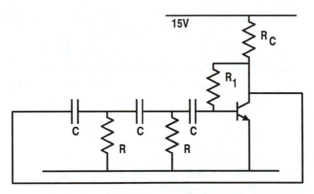

Figure 53.7: Example 53.1.

A suitable circuit is shown in Figure 53.7.

First set up the DC bias for the transistor. A reasonable value for a typical BC109 ($\beta = 300$) transistor amplifier would be $R_C = 4.7\,\mathrm{k}\Omega$.

We then require $V_C \approx 9\,\mathrm{V}$ so:

$$I_C = \frac{15 - 9}{4700} = 1.28\,\mathrm{mA}$$

$$\text{Then}\quad I_B = \frac{0.00128}{300} = 4.2\,\mu\mathrm{A}$$

$$\text{which gives}\quad R_B = \frac{9\,\mathrm{V} - 0.7\,\mathrm{V}}{4.2\,\mu\mathrm{A}} = 1.9\,\mathrm{M}\Omega$$

Given $f_0 = 300\,\mathrm{Hz}$ then:

$$300 \times 2 \times \sqrt{3}CR = 1$$

If we take $C = 0.01\,\mu\mathrm{F}$ we get $R = 980\,\Omega$.

53.2 Design a Wein bridge oscillator for a fixed frequency of oscillation of 2000 Hz using the circuit in Figure 53.6.

The essential equation is $f_0 = \frac{1}{2\pi CR}$.

Assume a reasonable value $C = 0.1\,\mu\mathrm{F}$, which gives $R = \frac{1}{2\pi f_0 C} = 796\,\Omega$.

Determine the value of R_B and then choose R_1 to get a loop gain of 1; that is, set the gain to at least:

$$A_V = 1 + \frac{R_1}{R_B} = 3$$

53.2 Problems

53.1 Show that the phase shift $\phi = 60°$ in a CR circuit when $\sqrt{3} = \frac{1}{2\pi f CR}$.

53.2 Sketch the voltage waveforms, showing the amplitude and phase, which you would expect to observe at the top of each of the resistors in the feedback CR network and at the base of the transistor in Figure 53.7.

53.3 Show that the centre of the band pass of the Wein filter in Figure 53.5 (a) is at a frequency $f_0 = \frac{1}{2\pi RC}$. Also show that the attenuation at this frequency is $\frac{1}{3}$ or $-9.5\,\mathrm{dB}$. (Review Problem 15.8.)

53.4 Design a transistor phase shift oscillator for a fixed frequency of 1500 Hz.

53.5 Design a Wein bridge oscillator for a fixed frequency of 700 Hz. Justify your choice of component values.

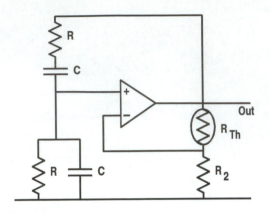

Figure 53.8: Problem 53.6.

53.6 Design a Wein bridge oscillator for a fixed frequency of oscillation of 1300 Hz using the circuit in Figure 53.8. Note that the type RA53 thermistor, R_{Th}, has a resistance of 5 kΩ at 20°C decreasing to 80 Ω at about 90°C.

53.7 The gain or sensitivity of the controller in a control loop, such as that shown in Figure 41.2, is gradually increased until the control loop oscillates with a periodic time, T. By how much does the gain of the controller have to be reduced so that the amplitudes of successive peaks resulting from a process disturbance are in the ratio of 4:1? What is the shortest time in which such a control loop can correct for a sudden disturbance?

53.8 In the circuit shown in Figure 53.9, the collector load is a resonant LC circuit. Calculate the frequency at which the gain will be a maximum. The secondary winding couples a fraction of the output signal back to the base. Under what circumstances will positive feedback occur? What will be the frequency of oscillation of the circuit? Under what circumstances will the output waveform be sinusoidal?

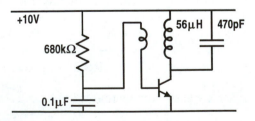

Figure 53.9: Problem 53.8.

Unit 54   Digital to analog conversion

In a digital to analog converter or DA converter:

- A resistive ladder is used to give successive fractions of a reference voltage.

- Each divided down voltage fraction drives a current, proportional to the voltage, through a resistor.

- The current through the resistors can be switched to the input of a current to voltage converter.

- The current to voltage converter sums all of the switched currents to generate a voltage output proportional to the binary value of the switches.

In analog electronics we deal with continuously variable quantities such as current or voltage which can have any value within a permissible range. These voltages or currents can then be used to represent continuous real world variables such as position, velocity, pressure, light intensity etc.

In digital electronics, quantities are represented by numbers which are themselves represented by discrete voltage levels. Since much of modern technology is based on the use of digital computers, it is necessary to have methods of obtaining an analog output from a digital computer (DA conversion) so that computers can exercise control over the external world and also it is necessary to have methods of obtaining a signal in digital form which is compatible with digital computers (AD conversion) and which represents the conditions in the continuously variable real world.

Digital computers carry out calculations in binary arithmetic, that is using a number system where the digits are 0s and 1s. The position of the digit is used to represent the significance of the number so that a conversion from binary to digital representation would be:

$$00101101_{Binary} = 1 \times 2^5 + 0 \times 2^4 + 1 \times 2^3 + 1 \times 2^2 + 0 \times 2^1 + 1 \times 2^0 = 45_{Decimal}$$

In order to represent this binary number in electronics, a bus, typically of eight wires in parallel, would be used in which a 0 V signal on the wire represents a logical 0 in the corresponding position of the binary number and a 5 V on the wire represents a logical 1 in the corresponding position of the binary number. The binary equivalent of 45 decimal, when placed on an 8 bit bus, would give the voltage configuration shown in Figure 54.1. The terms msb and lsb represent most and least significant bit.

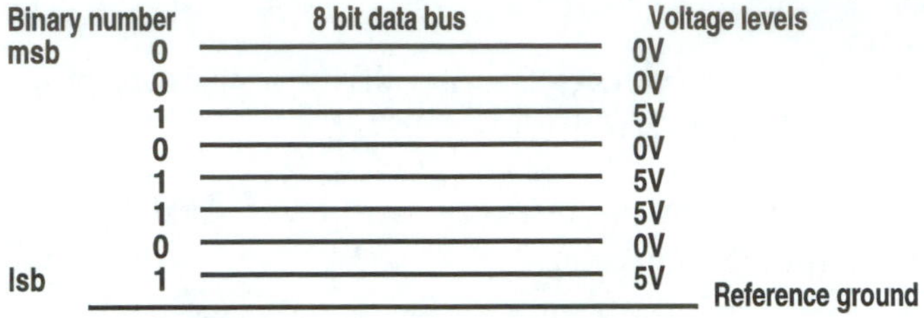

Binary number	8 bit data bus	Voltage levels
msb 0		0V
0		0V
1		5V
0		0V
1		5V
1		5V
0		0V
lsb 1		5V
		Reference ground

Figure 54.1: Binary number and corresponding voltage levels on the bus.

For simplicity, we will discuss digital to analog conversion in terms of 8 bit data buses but higher resolutions of 12, 14 and 16 bits are frequently used. In a DA conversion we convert the voltage levels on the eight wires of the bus into a single voltage level at a single analog output.

First, we set the range of the output. Typically this would be from 0 V to 1 V but it is straightforward to change the scale. The 8 data bits correspond to a range from 00000000 to a maximum of 11111111 in binary which corresponds to a range from 0 to 255 in decimal. So instead of an analog voltage which varies continuously from 0 V to 1 V we can only generate a set of 256 discrete values spaced at intervals of $\frac{1}{256} = 3.9\,\text{mV}$ apart. The resolution of the system is said to be 3.9 mV. Greater resolution is obtained by using more bits and a larger range of binary number.

The problem of D to A conversion has now been reduced to this: given eight wires going into a converter circuit, with voltage levels on these wires representing an 8 bit binary number, how can a voltage be generated at the output of the DA converter which is proportional to this binary number?

Many conversion systems have been developed but there is one system which is most commonly used called the R–$2R$ ladder converter.

Consider the circuit shown in Figure 54.2 (a). We have a reference voltage, V_{ref}, which is typically 1 V and is applied across a potential divider of R and two $2R$ resistors in parallel. This is equivalent to the potential divider of two equal resistors as shown in Figure 54.2 (b). The voltage at the centre of the potential divider is then $\frac{V_{ref}}{2}$. In Figure 54.2 (a), one of the $2R$ resistors is grounded at the bottom. The other $2R$ resistor goes to the switch and is either grounded or fed to the inverting input of a current to voltage converter circuit. Using the op-amp rules (Unit 39) the voltage at the inverting input is

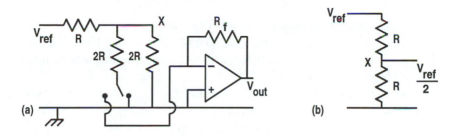

Figure 54.2: Single unit of R–$2R$ ladder network.

equal to the voltage at the noninverting input which is grounded. When the switch is in the left position, the bottom of the $2R$ resistor is held at ground voltage or $0\,\mathrm{V}$ by the operation of the op-amp. It is therefore at a virtual ground. The voltage at the mid point of the potential divider, marked X in the circuit, is therefore independent of the position of the switch and in this case the voltage is given by $\frac{V_{ref}}{2}$.

The current into the op-amp I–V converter is therefore the voltage at point X divided by the resistor to the switch, $2R$, which gives $I = \frac{V_X}{2R}$ and the output of the current to voltage converter is this current multiplied by the feedback resistor R_f. In this case V_X is half of the reference voltage V_{ref}, which therefore gives:

$$V_{out} = -\frac{V_{ref}}{2 \times 2R}R_f \quad \text{or} \quad 0$$

depending on the position of the switch.

At the point where V_{ref} is applied, the input resistance of the ladder is $R + R = 2R$ so if we put a copy of this elemental ladder into the circuit, where the right hand $2R$ termination resistor is located, we do not disturb or change the voltages in the circuit. This insertion can be repeated a number of times and the resulting circuit is shown in Figure 54.3.

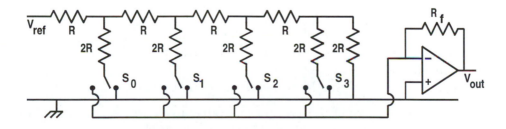

Figure 54.3: R–$2R$ ladder network.

The voltage at the equivalent of point X in the elemental ladder is reduced by a factor of 2 as we progress from left to right along the ladder. For clarity, the diagram only shows the circuit for 4 bits or stages of the ladder but it is easily extended to an 8 bit ladder.

The switches used in the ladder are not mechanical switches but are fabricated on the chip using enhancement mode MOSFETs connected to ground and controlled by the voltages on the relevant wire of the 8 bit data bus as shown in Figure 54.4 (a).

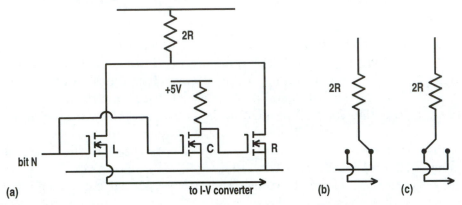

Figure 54.4: FET switches used with R–$2R$ ladder network.

If the signal on the data line for bit N is at $0\,\text{V}$ then FET L is OFF. FET C is also OFF so its drain voltage is at $+5\,\text{V}$ and FET R is ON so that the equivalent switch is as shown in Figure 54.4 (b).

If the signal on the data line for bit N is at $+5\,\text{V}$ then FET L is ON. FET C is also ON so its drain voltage is at $0\,\text{V}$ and FET R is OFF so that the equivalent switch is as shown in Figure 54.4 (c).

The current to voltage circuit in Figure 54.3 sums all of the currents through the $2R$ resistors from each of the voltage tap-off points on the R–$2R$ ladder and gives an output voltage:

$$V_{out} = -R_f \frac{V_{ref}}{4R} \left(S_0 + \frac{S_1}{2} + \frac{S_2}{4} + \frac{S_3}{8} + \ldots \right)$$

where S_N is the digital state of the relevant line on the data bus and takes the value 0 or 1.

The advantages of this R–$2R$ ladder network are that:

- Only two values of resistors (or one value if two are used in series) are required to be fabricated on the silicon chip.

- Typical values are $10\,\text{k}\Omega$.

- The fabrication can be carried out to the required precision if only one value is required.

- The resistors do not have to be precise values. They only need to be equal valued.

All of these features make this the most convenient DA converter for fabrication as an IC. For high precision it is possible individually to trim the resistors values on the chip by using a focused laser beam to cut out small sections of the resistive track. These high precision DA converters using laser trimmed resistors are more expensive and would only be used in critical applications.

54.1 Problems

In all of these problems you may assume that the resistive ladder is an $R-2R$ network of $10\,k\Omega$ and $20\,k\Omega$ resistors, as shown in Figure 54.3, and that the feedback resistor in the I to V converter is $10\,k\Omega$.

54.1 Calculate the output from an 8 bit DA converter for switch settings 01101010 when the reference voltage is $1\,V$.

54.2 Calculate the switch settings if an output of $0.63\,V$ is to be obtained from an 8 bit DA with a reference voltage of $4\,V$.

54.3 What is the smallest increment of output from an 8 bit converter having a reference voltage of $1\,V$?

54.4 What is the smallest increment of output from a 12 bit converter having a reference of $1\,V$?

Unit 55 Analog to digital conversion

In an analog to digital converter or AD converter:

- Feedback analog to digital converters include a DA converter.

- An algorithm is used to drive the DA converter which gives one of:

  - Ramp type conversion

  - Tracking conversion

  - Successive approximation

- Flash converters carry out the conversion at high speed using a resistor network and a bank of comparators.

- Integrating converters integrate the input voltage for a fixed time and then measure the time required to restore the output of the integrator to zero using a negative reference voltage.

There are many different analog to digital converters available. The Analog Devices *Data Conversion Products Databook* lists approximately 50 different families of AD converters. In all cases the device is used to obtain a digital value for an analog voltage presented at the input to the IC.

There are many features considered in the design of AD converters, such as accuracy, speed of conversion, price, compatibility with computer interfaces, linearity, precision, noise rejection, and it is compromises in the emphasis on these various features that lead to the wide range of available devices.

We will therefore not cover all of the available families of devices but rather look at the three broad groups of AD converters and see how they operate and try to see why certain families are better suited for certain applications.

Feedback analog to digital converters. This approach to AD conversion uses the digital to analog converters discussed in Unit 54 and incorporates them in a feedback system such as that shown in block diagram form in Figure 55.1.

A digital logic unit (DLU) generates an 8 bit binary number according to some algorithm. This number is fed to a digital to analog converter which

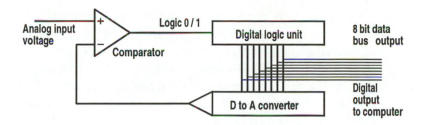

Figure 55.1: Block diagram of feedback AD converter.

generates the corresponding analog voltage and feeds it to a comparator. The comparator is basically an op-amp which is used without feedback and which has two output states: an output of 0 V or logical 0 when the inverting input is less than the noninverting input and an output of +5 V or logical 1 when the inverting input is greater than the noninverting input. The comparator output is fed to the DLU and controls the execution of the algorithm.

The system attempts to match the output of the digital to analog converter to the analog input voltage to within the resolution of the DA converter. The logical output of the comparator both determines the accuracy of the match and also controls the DLU in achieving a match. When a conversion is complete, the output of the DLU is also put on the 8 bit data bus output from which the data is read into the computer.

There are three basic types of algorithm or program which are programmed into the DLU by the manufacturer of the feedback AD.

A ramp type conversion algorithm is shown in Figure 55.2.

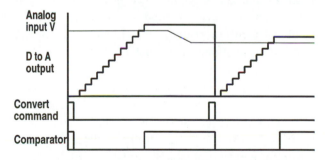

Figure 55.2: Ramp type feedback converter.

The DLU is programmed to operate as a counter which counts the cycles of a clock. The output of the counter is fed to the DA and the output of the DA therefore ramps up, or in fine detail moves up the voltage staircase, until the comparator changes state. At this point the clock input to the comparator is stopped and the count or data is read into the computer. In many cases the comparator output is also fed to the computer to signal the end of a conversion which signals that the data on the bus is now a valid representation of the AD conversion output and can be read by the computer. When the main computer determines that a reading is to be made, a conversion or measurement is initiated by putting a pulse on the convert command line as shown in Figure 55.2.

The second algorithm used in feedback AD conversion is the **tracking converter** for which the operating configuration is shown in Figure 55.3.

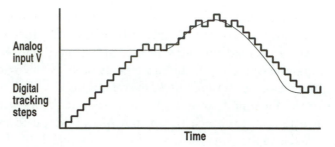

Figure 55.3: Tracking type feedback converter.

Initially this operates as a ramp type converter but it does not stop as soon as the point is reached where the comparator triggers, but rather the counter is switched from count-up mode to count-down mode so that the DA converter output continually brackets the analog input voltage. If the analog input voltage changes then the DA converter tracks it by taking more steps up than down on an increasing input voltage. The limitation of the system shows up when the analog input voltage changes at a rate faster than the maximum rate of the ramp or staircase as shown for a short segment of the falling edge of the sample shown in Figure 55.3.

Tracking type AD converters can also be used with 16 bit systems for digitizing audio signals in the preparation of masters for compact disks. The 16 bit digitizing resolution corresponds to the 100 dB dynamic range of hearing. The advantage of the tracking system is that it avoids the necessity for storing $44.1 \times 10^3 \times 16$ bits or 0.7 megabits of data per second of recorded sound when a sampling rate of 44.1 kHz is used.

Since the analog signal tends to be continuous, there is no point in going back to zero each time a conversion is required. Also, in audio signal digitization what is important is not the absolute pressure value on the microphone but rather the changes in the pressure, the sound, as recorded by the microphone.

In order to digitize an audio waveform, a sampling rate in excess of 20 kHz is required. If a 12 bit AD converter is used the data rate is then $20\,\text{kHz}\times 12 =$ 240000 bits per second which is greater than the available bandwidth in audio systems. In the tracking AD converter, only the changes need to be transmitted so the transmitted or recorded signal is a bit stream representing up or down steps. When conversion back to analog is required, a single bit DAC is used. A simplified representation of this converter is shown in Figure 55.5. A stream of logic 0 or 1 bits, representing the up and down steps from the tracking converter, are level shifted to $-5\,\text{V}$ and $+5\,\text{V}$ and applied to the gates of the two complementary MOSFETs so that the input to the integrator is connected to $-5\,\text{V}$ for a 0 bit or to $+5\,\text{V}$ for a 1 bit. The integrated and smoothed output then forms the analog signal for playback. The integration time is about $50\,\mu\text{s}$ to allow rapid tracking of the signal.

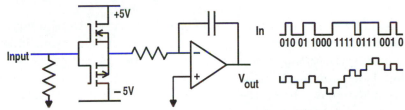

Figure 55.4: Principle of operation of single bit DAC.

The third type of feedback AD converter is the **successive approximation** type of converter. The strategy used in this conversion is the same as that used in the game where player A writes down a random number between 1 and 1000 and player B has to determine the number by guessing a value for the number and asking player A if the guessed value is the value written down. The most efficient strategy for B is to start with an initial guess of 500. If the reply is 'too low' then the next guess should be $500 + \frac{1}{2}\times 500 = 750$. If the reply is 'too high' then B takes away the increment and adds on half of the increment to guess $500 + \frac{1}{2}250 = 625$. At each stage the range of possible values is reduced by a factor of 2. This algorithm, as applied to a feedback AD converter, is shown in Figure 55.5.

The input signal is fed through a sample and hold amplifier to prevent the value changing while the conversion is being carried out. This sample and hold is not shown. The momentarily constant input voltage is shown as a fine line on the graph and marked Analog input V. Initially all 8 bits at the output of the DLU are set to 0. In the first time interval, the most significant bit of the DA converter is set to a 1. The result of the comparison is that $V_{in} > V_{DA}$ and therefore the 1 bit is left in position. In the next time interval the next most significant bit is set to 1 and the result of the

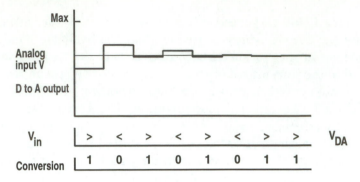

Figure 55.5: Successive approximation type feedback converter.

comparison is that $V_{in} < V_{DA}$ so that the bit is reset and a 0 is placed in the position.

Thus, in eight time intervals, it is possible to carry out an 8 bit conversion to an accuracy of 1 part in 2^8 or 1 part in 256. The minimum time interval is determined by the settling time of the DA converter and the comparator and may be of the order of $2\,\mu s$. So a typical AD conversion and data readout using this system can be carried out in about $20\,\mu s$ making this one of the faster types of AD converter.

A flash converter is shown in block diagram form in Figure 55.6.

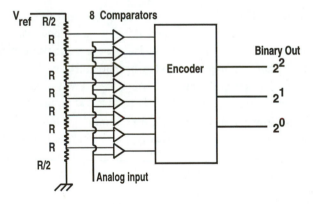

Figure 55.6: Flash type AD converter.

In this 3 bit converter, a resistive network consisting of seven resistors of value R (typically $10\,k\Omega$) and the top and bottom resistors of value $\frac{1}{2}R$ are used to give eight voltages over the range from $0\,V$ to the reference voltage, V_{ref}. The analog voltage input is simultaneously compared with each of these tapped off voltages in a bank of eight comparators. The logical outputs of these eight comparators are fed to an encoder which generates an output binary number which is the number of the lowest untriggered comparator.

The fact that the system operates in parallel on all comparators means that its speed is limited only by the comparator operating speed and therefore this system is used for very high speed operation such as digitizing television and video signals in real time. One line of a TV picture lasts 64 μs and requires about eight conversions per microsecond to give 500 pixels across the screen. The system shown in the figure is a 3 bit system of low resolution. An 8 bit system utilizes $2^8 = 256$ comparators and is a fairly substantial piece of electronics. It is normally installed as a card within a computer to minimize the time associated with the transfer of high speed data through the system or along cables. The complexity of the electronics is also matched by the cost of the converter and the cost of computer memory: 1MByte of memory is filled for each four digitized images at a 500×500 pixel resolution. Given that images are updated at a rate of 25 per second in a video system, a lot of expensive memory can be filled very quickly!

So in general, flash converters have a specialized range of application in digitizing very high speed events and in digitizing images which warrant the high system costs.

Integrating AD converters. These converters are widely used in digital multimeters. The basic block diagram of the converter is shown in Figure 55.7.

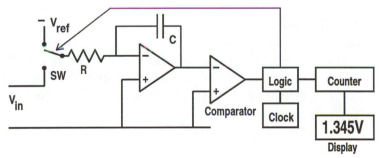

Figure 55.7: Block diagram of integrating type converter.

An input voltage is integrated for a fixed time in an op-amp integrator having a time constant CR. At the end of the time interval, T_{int}, the input voltage is disconnected from the integrator and a negative reference voltage is switched to the input to the integrator. This results in the output of the integrator returning towards zero volts. The time when this occurs is determined by a comparator and the time taken for the integrator output to be brought back to zero is recorded in clock cycles in a counter. Appropriate scaling of the R and C gives a count which is numerically equal to the input voltage.

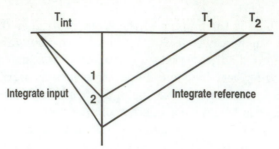

Figure 55.8: Operation of integrating type AD converter.

Figure 55.8 shows the voltages at the output of the integrator due to two different input voltages V_1 and V_2 being integrated for the fixed time interval T_{int}. The output voltage reached is greater for the larger input voltage V_2. When the switch is changed over to the reference, the rate of change of the integrator output is the same for the two measurements but T_2 is longer than T_1. The integrator therefore converts a voltage measurement to a time measurement: $T_n \propto V_n$. The values C and R are not critical. It is only required that they do not change during the short measurement time because they are used in both the integration of the input signal voltage and also in the integration of the negative reference voltage. The reference voltage does affect the accuracy of the scaling but this single component can be fabricated to the required high accuracy. The integration time is usually made to be equal to the period of the mains frequency so that any spurious pick-up of mains voltage will average to zero over a full cycle.

One chip normally contains the op-amp, comparator, clock and the liquid crystal display driver circuits and mass production gives the high accuracy, low cost hand held meters which are in use in all laboratories in large numbers.

55.1   Problems

55.1 What is the resolution in volts of a 12 bit AD converter with a range from $0\,\text{V}$ to $+5\,\text{V}$?

55.2 What is the binary output of an 8 bit converter with a range from $0\,\text{V}$ to $+10\,\text{V}$ for input voltages of $+1.3\,\text{V}$, $+4.3\,\text{V}$ and $+8.2\,\text{V}$?

55.3 What is the average conversion time of an 8 bit ramp type converter in which a $10\,\text{kHz}$ clock is used?

55.4 Draw a flow chart for a program to control the operation of the successive approximation feedback converter shown in Figure 55.5.

Unit 56   SCRs and triacs for power control

- A silicon controlled rectifier (SCR) can be triggered into conduction in the forward direction by application of a pulse to the gate terminal of the SCR.

- A triac can be triggered into conduction in either direction by application of a pulse to the gate terminal.

- A unijunction transistor (UJT) is used in a relaxation oscillator to give a train of trigger pulses for SCRs and triacs.

- A diac (bidirectional diode thyristor) conducts when the threshold voltage of about $20\,\mathrm{V}$ is exceeded. Once triggered, the device resistance then drops rapidly.

- In phase angle triggering of triacs, a variable segment of each half waveform of current is passed through the load.

- In burst fire control, the full power is applied to the load for a variable fraction of the time to get the required average power dissipation. The ON time corresponds to an integral number of half cycles of current.

When a transistor is used to control the current through a load, the full voltage of the power supply is applied across the transistor and load in series to give $V_S = V_T + V_L$. The same current, I, flows in both the transistor and the load so the power dissipated in the transistor is $I \times V_T$ and the power dissipated in the load is $I \times V_L$. This is not important for small currents because a simple heat sink is sufficient to dissipate the heat from the transistor but for large currents of the order of amps and for voltages of the order of hundreds of volts the power dissipated in the transistor is usually too great to be readily dissipated and heat damage to the transistor usually results.

If the system is of necessity linear, such as in the case of high power audio amplifiers, radio transmitters etc., then the older technology of thermionic valves may have to be utilized. However, if the requirement is that the average power or current in the load be controlled then rapidly switching the full current on and off to get an average current somewhere between the minimum and maximum can be very effective. There is therefore a need for an electronic switch which can turn current on and off very rapidly without switch wear and which can be used for controlling large currents.

No power is dissipated in an ideal switch because when the switch is open $P = V \times I = V \times 0 = 0$ since there is no current and when the switch is closed again $P = V \times I = 0 \times I = 0$ because the closed switch has zero resistance and there is no voltage drop across the switch.

One problem with switches is that they are mechanical and slow, typically taking a fifth of a second to operate. But a more significant problem is that each time a switch operates, there is an arc or spark at the contacts which leads to erosion of the metal and in extreme cases welding together of the contacts. One example of contact wear is the need to replace the points in a car ignition system every 15,000 km or sooner because of wear of the contact metal. In the car, the function of the points is to interrupt the current in the primary of the high tension coil or transformer which generates the spark for ignition of the petrol.

In principle a transistor can be switched or driven rapidly from the non-conducting state to the conducting state and made to spend a very short time in the high power dissipation zone, but the necessary circuitry is complex. A more efficient approach is to use a latch circuit in which positive feedback is employed to give very rapid switching.

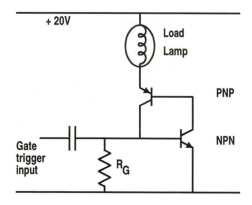

Figure 56.1: A two transistor latch circuit.

Figure 56.1 shows a two transistor latch circuit. When the supply voltage is turned on, there is no current in either transistor and therefore there is no current in the lamp which is the load in this example. The lamp could be replaced by a motor or a heater but a filament bulb is very convenient for experiments in the laboratory and also for testing circuits in the field.

The circuit operates as follows. With the latch in the OFF state, the R_G resistor, typically $10\,\text{k}\Omega$, holds the base of the npn transistor at the emitter voltage so there is no base current and therefore there is no collector current in the npn transistor. This means that there is no base current for the pnp transistor which means that there is no collector current in the pnp transistor which gives no base current in the npn transistor as we assumed at the beginning. So the circuit configuration is logically consistent with no current in either transistor. (Note the logical argument which is used here and which is often very useful in electronics: make an assumption about the input and trace the effects around the feedback loop back to the input; if the loop back is consistent with the assumption made about the input then that is a stable configuration for the loop.)

Now apply a short positive pulse of about a volt to the input through the capacitor. This causes the npn transistor to become forward biased and the transistor conducts momentarily. The collector current of the npn transistor is the base current of the pnp transistor so the pnp transistor conducts and this gives collector current in the pnp which gives base current in the npn which reinforces the initial trigger pulse from the external source. The current is amplified by a factor of about 10,000 each time the signal propagates round the loop so there is very rapid switching from the OFF state to the ON state for both transistors.

The current continues to grow until it is limited by the resistance, R_L, of the load. The current growth then stops and the circuit enters a stable state with about $1\,\text{V}$ across each transistor and both transistors conducting. The latch circuit is then stable in the ON state and will remain in this ON state until the power is switched off.

In principle, a large negative pulse at the trigger input could turn off the latch but in practice it is found that it is difficult to turn off a latch circuit once it has latched on, basically because the turn-off trigger pulse must supply all of the current through the latch for long enough for the latch to stop conducting or 'drop out'. This is therefore one difference with a mechanical light switch which latches equally well to the OFF state as it latches to the ON state. An electronic latch turns on easily but is difficult to turn off.

In Figure 56.2 (a) we show the two transistor latch which we have just discussed. The connections between the npn-pnp transistor layers are shown in Figure 56.2 (b). In Figure 56.2 (c) we have merged the connected layers to get a four layer device which is called an SCR or silicon controlled rectifier or thyristor and for which the conventional circuit diagram symbol is shown in Figure 56.2 (d).

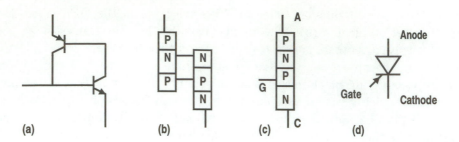

Figure 56.2: Evolution of a latch circuit to an SCR.

The performance of the SCR is exactly the same as that of the two transistor latch which we have discussed except that the device is available with voltage ratings up to 1000 V and current ratings up to a few hundred amps.

In the operation of SCRs there are a few important parameters which must be considered:

- SCR has the word 'rectifier' in its name. Current can only flow through the SCR in the direction of the arrow in the circuit symbol of the diode. Current then only flows when the SCR has been triggered.

- A holding current or minimum current must flow through the SCR in order to maintain it in the conducting state.

- Break over voltage is the maximum voltage which can be applied between the anode and the cathode of the SCR in the forward direction without the SCR going into the conducting state in the absence of a trigger to the gate.

- V_{RRM} is the maximum reverse blocking voltage of the SCR.

- Trigger voltages and currents are typically of the order of 2 V and 20 mA for about 5 μs.

- $\frac{dV}{dt}_{max}$ is the maximum rate of change of the anode voltage which will not cause spurious triggering of the SCR.

- $I_{T\,ave}$ is the maximum average current rating of the SCR.

In general, it is necessary that attention be paid to the maximum ratings of the SCR because the applications are usually at higher voltage and higher current than normal transistor applications and the possibility of serious heat damage to the SCR is that much greater.

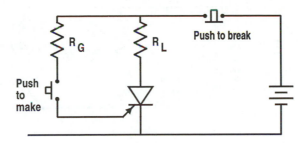

Figure 56.3: DC triggering of an SCR.

We now examine some representative circuits which show the operation of SCRs and demonstrate the basic principles of SCR triggering. The simplest trigger circuit is the DC circuit shown in Figure 56.3 where switches are used to control the small trigger signals to the SCR. The resistor, R_G, limits the gate current to the permissible limit. For instance, if the specified gate current is $I_G = 10$ mA and with a 20 V supply an $R_G = \frac{20}{0.01} = 2$ kΩ would give reliable triggering when the push to make triggering button is pressed. The load, R_L, could be any load which draws a current less than the maximum average current for the particular SCR, $I_{T\,ave}$.

Some care must be taken when the load is a filament bulb since the current which flows immediately after the bulb is turned on can be up to five times the current which flows when the bulb is hot, owing to the change in the resistance of the filament as it heats up, and therefore you must choose an SCR with the current rating for the higher starting current. If the load is a motor with a commutator then there is the possibility that a momentary interruption of the current due to a bad contact at the commutator can cause the SCR to drop out of conduction.

In the circuit in Figure 56.3 the current can only be turned off by interrupting the current through the SCR by pressing the push to break OFF button so this circuit is really only useful for demonstrating the principle of DC SCR triggering.

It is when the SCR is used in AC circuits that the versatile aspects of SCRs start to become apparent. Figure 56.4 shows an SCR circuit powered from an AC supply. For safety reasons, in laboratory experimental work, you should use an isolating transformer. The circuit will, however, operate satisfactorily and switch mains voltages without the use of any isolation. But you cannot connect oscilloscopes to the mains without extreme care in the use of the ground connection of the oscilloscope test lead so the opportunities for diagnostic exploration are restricted.

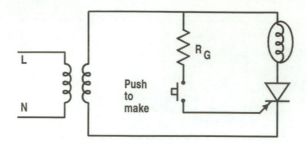

Figure 56.4: Triggering of an SCR in an AC circuit.

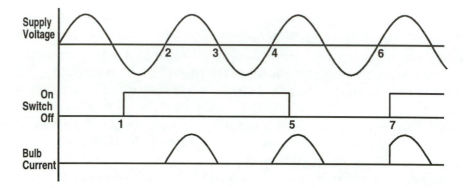

Figure 56.5: V and I waveforms in an AC SCR circuit.

A waveform diagram for this circuit is shown in Figure 56.5 which shows up a number of important features of SCRs when used with AC triggering. The relevant points in the waveforms in Figure 56.5 have been labelled with numbers for the purposes of this discussion.

1. The ON switch is pressed but no current flows because the SCR is reverse biased.

2. The waveform goes positive which gives positive gate current and positive voltage across the SCR so the SCR triggers and current flows in the load as shown in the graph of bulb current, I_{Bulb}.

3. The waveform goes negative and the SCR stops conducting because of the rectifier or diode action of the SCR.

4. The voltage goes positive again and the button is still pressed so the SCR conducts.

5. The trigger button is released but the SCR continues to conduct, because of the latching action of the SCR, until the end of the half cycle when the SCR drops out of conduction.

6. The SCR is OFF because there is no gate trigger.

7. The trigger button is pressed during a positive half cycle and the SCR immediately conducts for the remainder of the half cycle.

Triggering of SCRs by applying DC or AC signal switching to the gate does operate but has one great disadvantage in that there is a wired or galvanic connection between the trigger circuit and the high voltages and currents which are present in the power circuit controlled by the SCR. This can constitute a significant hazard of electrical shock for operators.

In an effort to provide isolation between the sensing and control circuits and the power circuits a device called a unijunction transistor or UJT has been developed. Figure 56.6 (a) shows a schematic diagram of the construction of a UJT.

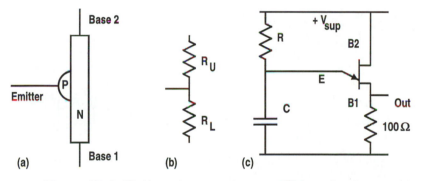

Figure 56.6: Unijunction transistor oscillator circuit.

The device is essentially a long resistor of n-type silicon. The n-type doping concentration is low and therefore the resistance of the resistor is high. At a point about half way along a p-type junction is made to the n-type resistor which gives a point pn junction. When a voltage is applied between base 1 and base 2, the two ends of the resistor, the device behaves just like a potential divider with the voltage at the p-type junction given by the usual potential divider formula. Since $R_U \approx R_L$ initially, the potential at the junction is about half way between zero and the supply voltage. When the external voltage applied to the emitter terminal is less than the voltage set by the effective potential, the pn junction is reverse biased and no current flows through the junction.

However, if the voltage applied at the emitter rises above the voltage at the centre of the bar as set by the potential divider then the pn junction becomes forward biased and p-type holes are injected into the bar from the heavily doped p-type region. These injected holes move down into the lower resistor, R_L, and lower the value of R_L drastically from a typical value of $\approx 50\,\text{k}\Omega$ to a value $\approx 10\,\Omega$.

The device is used in a circuit such as that in Figure 56.6 (c). The capacitor charges exponentially from zero towards the supply voltage and eventually reaches the trigger voltage for the device which is given by ηV_{sup} where $\eta \approx 0.6$ is called the intrinsic stand-off ratio for the UJT. At this point the resistance from the emitter to the base 1 terminal decreases rapidly owing to the injection of holes and the charge on the capacitor flows through the $100\,\Omega$ resistor to discharge the capacitor and to give a very fast (typically $5\,\mu\text{s}$) current pulse at the base 1 terminal. When the capacitor discharges, the injection of holes into the lower resistor stops and the resistance, R_L, reverts to its original high value so that the charge-up of the capacitor starts again giving a continuous train of pulses at base 1.

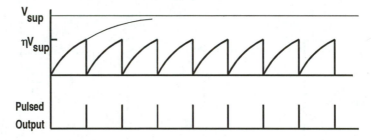

Figure 56.7: Voltage waveforms in a UJT oscillator circuit.

This is shown in Figure 56.7 where the charging of the capacitor towards the supply voltage is shown together with the fast pulses at base 1. If you build this circuit, you will have to adjust the oscilloscope trigger carefully if you are to see these fast pulses clearly.

The period of these pulses is approximately given by $T \approx RC$ with R in the range from $5\,\text{k}\Omega$ to $30\,\text{k}\Omega$ for typical UJTs such as the 2N2646 device.

A typical application of the device is shown in Figure 56.8 where the level of water in a tank is sensed by the float level switch which closes when the tank is full. For a partially empty tank, the level switch is open so the UJT circuit is free to oscillate and deliver a train of pulses at base 1 terminal. A small transformer, called a pulse transformer, is used to couple the UJT pulse to the gate of the SCR. The pulse transformer has very high primary to secondary voltage isolation and is designed for operation with high frequency

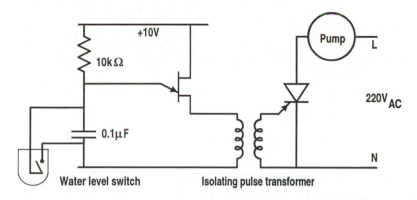

Figure 56.8: Use of UJT oscillator for galvanic isolation.

pulses giving good isolation from electrical shock between the sensing circuit and the power circuit. The function and operation of this transformer is very similar to that of the transformer used in the isolation amplifiers discussed in Unit 51.

The rapid train of pulses at the gate of the SCR triggers the SCR allowing DC to flow through the DC motor in the pump which then fills the tank. When the level in the tank rises to the level switch, the switch closes and shorts out the capacitor of the UJT oscillator, stopping the oscillation and the train of pulses to the SCR. The current in the motor then stops at the next zero crossing of the mains voltage waveform so that the tank does not overfill. If water is drawn off from the tank, the level drops, the switch opens, the UJT oscillator starts again and supplies triggering pulses to the SCR so that water is pumped in again to maintain the level.

The circuit in Figure 56.8 illustrates another point about SCRs. When they are used with mains supplies in control applications, the SCR is usually located on the neutral side of the load heater or motor whereas a normal switch or contactor is usually located in the line from the live of the mains supply. With the SCR on the neutral side, the motor is at mains supply voltage when it is off and there is therefore a great danger of electrical shock for anyone who is not careful!

SCRs allow the control of DC or of rectified AC but do not pass the full AC waveform and so cannot be used with most induction motors.

Full wave control could be achieved by using two SCRs back to back as shown in Figure 56.9 but a better approach is to use a triac which is equivalent to two back to back SCRs with a single gate terminal as shown in Figure 56.9.

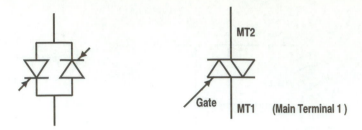

Figure 56.9: Evolution of two SCRs to a triac.

Use of a triac allows the implementation of phase control of triggering as shown in Figure 56.10.

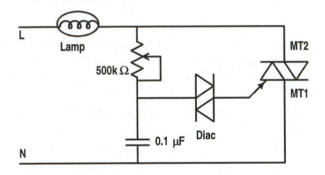

Figure 56.10: Phase angle triggering of a triac.

The variable $500\,\mathrm{k\Omega}$ resistor and the $0.1\,\mu\mathrm{F}$ capacitor allow a variable delay to be generated between the zero crossing of the mains voltage waveform and the time when a trigger pulse is applied to the triac. A device called a diac is used in this circuit. The diac does not conduct until the voltage across it reaches about $20\,\mathrm{V}$. It then conducts and the resistance decreases rapidly so that, in the circuit in Figure 56.10, the diac dumps all of the charge stored in the capacitor into the gate of the triac and causes it to trigger. The effect of the CR and diac is to delay the application of the trigger pulse until some set time after zero-crossing.

The resulting current flow in the lamp is shown in the diagram in Figure 56.11. As the value of the variable resistor is reduced, the voltage on the capacitor reaches the diac trigger voltage earlier and the triac is triggered earlier in the waveform giving nearly continuous variability in the intensity of the lamp. These circuits are available for use as dimmer switches. The circuit, as drawn in Figure 56.10, will generate substantial interference due to the sharp pulses at switch-on (high frequency Fourier components) which can

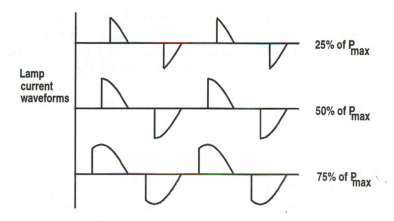

Figure 56.11: Waveforms in phase angle triggering circuit.

cause problems in radio reception and also cause problems of waveform distortion and harmonic loading for the power supply company. A commercial dimmer switch will have extra suppressor circuitry included which eliminates this interference problem.

When the electrical power supply to heaters, which have a longer time constant than lamps, is to be controlled, a better system is to use a zero volt switch system. These switches are available as made-up components containing all of the circuitry and the power control triacs in one package.

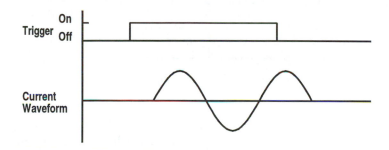

Figure 56.12: Waveforms in a zero volt switching controller.

In the operation of a zero volts switch, a DC signal current is applied at the control input which passes a current through an internal light emitting diode (led). The light is detected with a photodiode so that there is opto-coupled isolation between the input control signal and the power circuits. There are other internal circuits which are designed to block the application of trigger pulses to the gate of the triac until the zero crossing of the mains waveform. This causes the current to switch-on when the voltage is low and therefore the current surge at switch-on is minimized and the interference is also minimized. The result of this triggering strategy is that a whole number of half cycles of the mains flow through the load and the waveforms are as shown in Figure 56.12.

56.1 Problems

56.1 Calculate component values for a UJT metronome circuit in which the $100\,\Omega$ resistor in Figure 56.6 (c) is replaced by a loudspeaker and which uses a variable resistor in place of R to obtain a pulse repetition rate from 1 per 3 seconds to 10 per second.

56.2 Calculate the power dissipation in a load resistor as a percentage of maximum for triggering phase angles of $10°$, $30°$, $120°$, $135°$ and $160°$.

56.3 A $220\,\mathrm{V_{RMS}}$ waveform is applied to an RC filter of $120\,\mathrm{k\Omega}$ and $0.1\,\mu\mathrm{F}$. Calculate the time delay between the zero crossing of the driving voltage waveform and the time at which the voltage across the capacitor crosses $32\,\mathrm{V}$. Will the time delay be different if the voltage across the capacitor is zero at the time of the zero crossing of the driving voltage waveform?

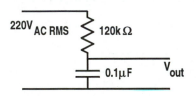

Figure 56.13: Problem 56.3.

56.4 A UJT oscillator, operating at a frequency of $100.2\,\mathrm{Hz}$, is used to provide trigger pulses to the gate of a triac which controls the current in a tungsten filament bulb powered from a $50\,\mathrm{Hz}$ mains supply. Describe the variation of the light intensity from the bulb and justify your answer.

Unit 57   Nonlinear circuits and chaos

- Chaotic systems are characterized by a very sensitive dependence on the initial conditions which causes initially adjacent system states to diverge exponentially.

- A chaotic system must have at least three degrees of freedom if the system state is never to repeat.

- Chaotic systems may start in a stable state but become periodic and then chaotic as some parameter is varied by following a route to chaos. A common route to chaos is associated with period doubling.

- Many chaotic systems enter a stable, nonrepeating chaotic orbit which when projected onto a plane forms a pattern called a strange attractor.

We have examined linear systems in some detail since such linear systems are the most commonly used in electronics. Even when the system becomes mildly nonlinear, such as is shown in the example of saturation illustrated in Figure 53.3, the system still remains stable and predictable. What we will examine in this unit are the characteristics of nonlinear circuits which become chaotic but calculable and are described under the heading of deterministic chaos.

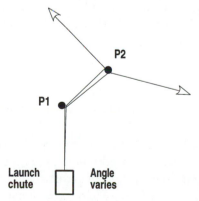

Figure 57.1: Pinball machine.

The concept of a deterministic chaotic system whose eventual state exhibits a sensitive dependence of initial conditions is well illustrated by a simple pinball mechanical system as shown in Figure 57.1.

A ball is launched at the array of pegs. It strikes the first and recoils at an angle which depends strongly on the impact direction. It then impacts on a second pin again having an angle of recoil depending on the line of incidence. The final position of the ball is plotted after a fixed time. As time increases the position of the ball becomes more and more sensitive to the initial trajectory of the ball. The final position in a chaotic system becomes an exponentially diverging function of the number of recoils off the pegs. Or to put it differently, two initially adjacent trajectories will diverge to give an eventual position which could be anywhere in the available space.

In the system in Figure 57.1, we can use only one peg and plot the position of impact of the ball on the border as a function of the angle of launch. The distribution of impact points is calculable. Put in a second peg and repeat the calculation or the experiment. Some of the trajectories such as those where the ball does not strike more than one peg are easily calculated. As the number of pegs is increased the distribution of impact points becomes more and more random or chaotic.

This example has illustrated one of the properties of chaotic systems, an exponential divergence characterized by what is called a Lyanopov exponent.

Another feature of some chaotic systems is that the systems are initially stable but as some parameter is varied they become chaotic. There is what is called a route to chaos, the most common of which is the period doubling route to chaos.

Consider a population of frogs living in a pond. In the simplest model, the population x_{n+1} in any one year depends on the population x_n in the previous year and the reproduction rate, r, so that:

$$x_{n+1} = r \times x_n$$

It is easily seen that the frog population is only stable and constant when $r = 1$. If $r > 1$ there is a population explosion and if $r < 1$ the frogs become extinct.

However, frogs eat flies and if there are more frogs there will be less flies, so we can visualize an improved model of the population dynamics in which we describe the number of flies available for food by a term $(1 - x_n)$ where $x_n = 0$ corresponds to no frogs and $x_n = 1$ corresponds to the pond full of frogs! Now let us change the population growth model and assume that the number of frogs in the next generation depends on the reproduction rate, r, the number of frogs, x_n, and the available food (flies), described by $(1 - x_n)$.

The number of frogs in the next generation is then given by the logistic equation:

$$x_{n+1} = r \times x_n \times (1 - x_n)$$

In order to examine the behaviour of the population we carry out a numerical model of the system. This is an operation which is very frequently performed in modelling chaotic systems. In dynamical systems it involves numerical modelling of the differential equations which describe the system and frequently requires significant computing power. In this case, however, we will use a straightforward iteration to calculate a population from the parameters and the previous year's population.

The logistic equation $x_{n+1} = r \times x_n \times (1 - x_n)$ should be programmed into a computer and run for about 1000 iterations for various values of the parameter r. When the last 20 values of x_n are printed to the screen and are are examined it is found that for any initial nonzero value of x_0 and for a range of values for r, number sequences similar to those shown in the table will obtained.

$r =$	1.0	1.5	2.0	2.5	3.0	3.5	3.55
1001	0	0.33	0.5	0.6	0.674	0.501	0.540
1002	0	0.33	0.5	0.6	0.659	0.874	0.882
1003	0	0.33	0.5	0.6	0.674	0.383	0.370
1004	0	0.33	0.5	0.6	0.659	0.826	0.827
1005	0	0.33	0.5	0.6	0.674	0.501	0.506
1006	0	0.33	0.5	0.6	0.659	0.874	0.887
1007	0	0.33	0.5	0.6	0.674	0.383	0.354
1008	0	0.33	0.5	0.6	0.659	0.826	0.812
1009	0	0.33	0.5	0.6	0.674	0.501	0.540
1010	0	0.33	0.5	0.6	0.659	0.874	0.882

This program is run for values of r extending from 0 to 4 and the resulting values of x are plotted on a graph as shown in Figure 57.2. From the table there is only one value, $x = 0.6$, for $r = 2.5$ but there are four values of x for $r = 3.5$. At each bifurcation point on the graph the number of iterations which are required before the pattern repeats increases by a factor of 2 which is why this is named the 'Period doubling route to chaos'. For values of r greater than 4 the pattern does not repeat and the succession of numbers is termed chaotic. This sequence of numbers for $r = 4$ is not a set of random numbers, however, because each number is computed or determined from the previous number and the system is therefore said to exhibit deterministic chaos.

So we now have a system which may be stable for a certain range of a parameter but which then becomes more and more unstable along a period

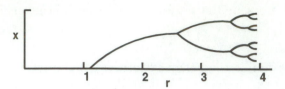

Figure 57.2: Period doubling route to chaos.

doubling route as the parameter is varied away from the stable region. An important distinction should be made at this point. In an electronic system distortion is characterized by higher Fourier harmonics appearing in the output but this is not chaotic behaviour. Period doubling is characterized by subharmonics appearing in the output and this is a signature of the imminence of chaos.

Linear systems do not exhibit chaotic behaviour. Nonlinear systems usually introduce higher frequency Fourier components and occasionally exhibit chaotic behaviour but, until recently, systems which exhibit chaotic behaviour were either not understood or not recognized and therefore tended not to be discussed in textbooks. Any known but unrecognized cases of chaotic behaviour were regarded as bad engineering and avoided. Now that we can recognize, describe and understand the phenomenon of chaos there is a trend towards utilizing chaotic systems for useful purposes.

A number of electronic circuits have been proposed as examples of circuits which exhibit chaotic behaviour but in some cases, dating from the early days of the subject, electronic reality did not correspond to the mathematical models which purported to show and explain the chaotic behaviour.

We will restrict our discussion of chaotic circuits to one of the better characterized chaotic circuits which was initially developed by Leon Chua and which is now named after him.

In this treatment of chaotic electronic systems we take a simple system and attempt to follow through the analysis of the system from the viewpoint of electronics and not from the more abstract mathematical viewpoint. This approach to understanding the simpler case should then be applicable, by extension, to more complex chaotic electronic systems. Chaotic systems are nonlinear and the usual approach of using differential equations in the analysis will not work so numerical methods are employed. This leads to a loss of intuitive understanding of the circuits owing to the intervening layer of programming and modelling.

In carrying out the analysis, the principle of Occam's Razor will be applied. While the principle is ascribed to William of Occam, it does not actually appear stated in his works. It is that 'Entities are not to be multiplied without necessity' or in more modern phrasing 'Keep it simple'. Unfortunately owing to the complex nature of chaotic systems, the explanations are not always simple but we will endeavour to keep the flow of the argument as straightforward as possible.

The fundamental unit used in this circuit is a device called a Chua diode. It is essentially a nonlinear negative resistance. A passive negative resistance is impossible since a negative resistance is essentially a device which outputs electrical power, so there must be an energy source somewhere in the system. In our case the energy source is the power supply for the op-amp. However, we have already met a negative resistance when we discussed the photodiode. The lower right hand quadrant of the I–V characteristic for a photodiode shown again in Figure 57.3 is a region where the voltage across the device is positive but the current is negative thus giving a negative resistance. The light energy falling on the photodiode is being converted into electrical energy and is driving the external circuit. By a similar argument a battery or a dynamo can also be considered as a negative resistance device.

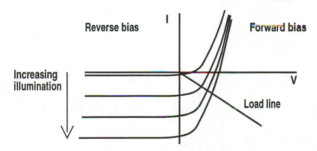

Figure 57.3: Photodiode characteristic.

A circuit which synthesizes a Chua diode is shown in Figure 57.4. We start with two diodes having resistors in series with them as shown in Figure 57.4 (a). The diodes do not conduct until the knee voltage of 0.7 V is reached and after that, the current is limited by the 3.3 kΩ series resistor so the I–V characteristic is as shown.

The second part of the circuit is shown in Figure 57.4 (b). In analyzing the operation of this circuit we use the rule that the voltage difference between the input terminals of the op-amp is zero. The voltage across the 1.1 kΩ resistor is therefore equal to the voltage at the top of the circuit. The output voltage from the op-amp must then be sufficient to drive current through the 300 Ω and 1.1 kΩ in series to give this voltage. If a voltage of 2 V is present at the top of the circuit then the op-amp output is $\frac{1100+300}{1100} \times 2 = 2.54$ V. This op-amp output is also applied to the upper 300 Ω resistor and gives a current of $\frac{2-2.54}{300} = -1.8\,\text{mA}$ in the input corresponding to an input resistance of $-1.1\,\text{k}\Omega$.

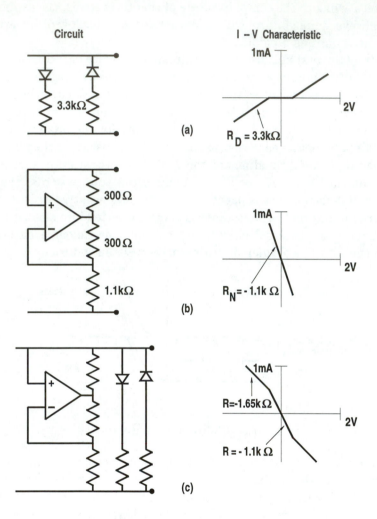

Figure 57.4: The operation of Chua's diode.

When these two circuits are placed in parallel as shown in Figure 57.4 (c) the central region has a resistance of $-1.1\,\text{k}\Omega$ but the outer regions have a resistance of $-1.65\,\text{k}\Omega$ corresponding to $+3.3\,\text{k}\Omega$ in parallel with $-1.1\,\text{k}\Omega$.

In combining the characteristics for each of the circuit segments it should be noted that since the two circuits are in parallel, we must add the currents at each voltage to get a resultant. When you are experimenting with this circuit, you can observe these characteristics by using a curve tracer which applies an alternating drive voltage to a circuit and measures the resulting current. The voltage is displayed on an oscilloscope X axis and the current is displayed on the Y axis as in the I–V characteristics. Some oscilloscopes have a component test facility which does the same thing but with less control over the voltages which can be applied. Alternatively you could use a circuit similar to that in Figure 17.4 where a function generator is used to drive the circuit (in place of R) and the capacitor is replaced by a low valued resistor ($10\,\Omega$) which is used to sense the current. The oscilloscope is then used in XY mode.

Now put a variable resistance and a parallel LC in series with this Chua diode circuit as shown in Figure 57.5. The variable resistance should be adjusted until oscillation is obtained. If the resistance of the variable resistor is then measured (out of circuit) it will be found that the value which gives stable oscillation is between the resistance of the two segments of the Chua diode ($1.1\,k\Omega$ and $1.65\,k\Omega$ in this example). The oscillation is driven by the negative damping effect of the negative resistance of the Chua diode. The amplitude does not increase indefinitely because the series resistance of the Chua diode and the R_V becomes positive for larger amplitudes of oscillation thus limiting the amplitude.

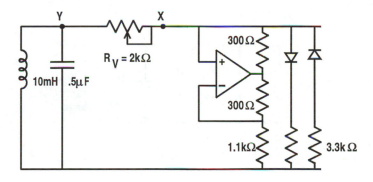

Figure 57.5: Chua's oscillator circuit.

This circuit oscillates with an approximately sinusoidal waveform at a frequency given by $f = \frac{1}{2\pi\sqrt{LC}}$ as shown in the oscilloscope tracing in Figure 57.6 of the waveforms at points X and Y.

If a capacitor is added to the circuit between point X and ground to give the circuit shown in Figure 57.7, then a circuit is formed which exhibits chaotic oscillation. In the circuit in Figure 57.7 we have replaced the op-amp and two diodes forming the Chua diode by the boxed resistor symbol which is now the more usual representation of the Chua diode. This is appropriate

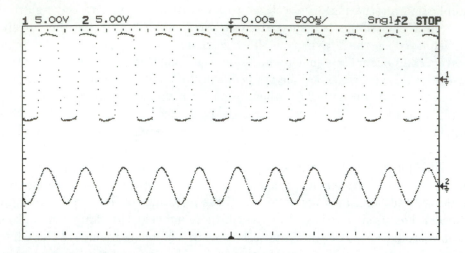

Figure 57.6: Storage oscilloscope printout of Chua oscillator waveforms.

since the nonlinear negative resistance diode is now available as an integrated circuit component.

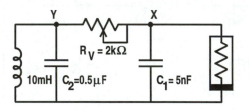

Figure 57.7: Chua's chaotic oscillator.

After the variable resistor R_V has been adjusted with some delicacy you should get the oscilloscope waveforms similar to those shown in Figure 57.8 with the signals being taken from points X and Y in the circuit. If you fail to obtain these waveforms then you may have to substitute slightly different values for some of the components. It has been found that the inductor is a critical component. If the resistance of the inductor is too large the circuit will not operate. Also you might try substituting slightly different values for the 1.1 kΩ shown in Figure 57.5. The difficulty is that the operation of this circuit depends on differences of about 1% between two resistors which are only specified to a 5% tolerance. One solution is to use a variable resistor instead of the 1.1 kΩ resistor and trim the circuit until it operates.

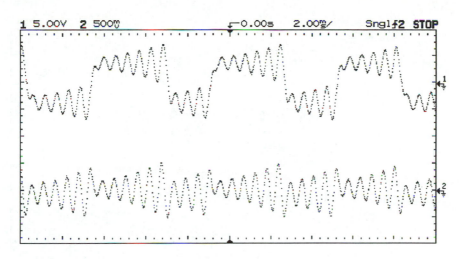

Figure 57.8: Voltage waveforms measured at points X and Y of Figure 57.7.

Another form of presentation of the performance of a chaotic oscillator circuit is shown in Figure 57.9 where the oscilloscope is operated in XY mode with the same signals as those in Figure 57.8. This representation shows what is called the double scroll chaotic attractor for the circuit.

We will now present an explanation in electronic terms of the operation of the chaotic oscillator circuit. A full mathematical treatment of the operation is available in the literature (see, for instance, Kennedy, M.P., *IEEE Transactions on Circuits and Systems*, **40**, (10), 640, 1993) but the mathematical model is not necessarily the best way of obtaining a first understanding of the operation of the circuit.

This explanation is presented essentially as a superposition of three distinct mechanisms which operate in the circuit.

First consider a potential divider of R_V and R_N where R_N is the Chua diode nonlinear negative resistance as shown in Figure 57.4 (c).

If a voltage is applied to this potential divider, as in Figure 57.10, the output voltage will be:

$$V_{out} = \frac{R_N}{R_V + R_N} \times V_{in}$$

But R_N is negative so when R_V is slightly less than the magnitude of R_N we will obtain an output voltage from the potential divider which is **larger** than the input voltage. The voltage at point Y across the LC in Figure 57.7 and shown in the lower trace in Figure 57.8 can be considered as the input to the potential divider and the voltage at point X in Figure 57.7, shown as the

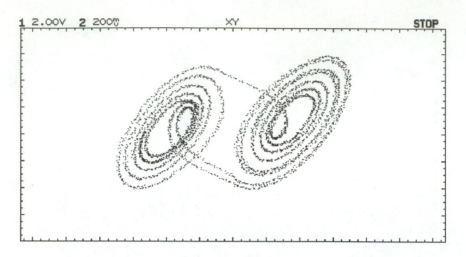

Figure 57.9: Double scroll chaotic strange attractor.

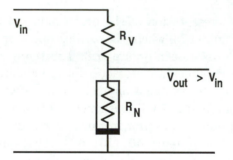

Figure 57.10: Negative resistance amplifier.

upper trace in Figure 57.8, can be considered as the output. Examination of the two traces in Figure 57.8 does show that there is an amplification and that the output of the potential divider is indeed greater than the input. The sensitivity settings of the oscilloscope are shown at the top left of the display.

Next, consider the series circuit of the R_V and the R_N of the Chua diode. Since these are in series we combine the I–V characteristic curves by adding the voltages horizontally for each current to give the resulting characteristic as shown in Figure 57.11.

This characteristic has an unstable equilibrium point at the origin and if an LC is connected across the input, the negative resistance of the I–V characteristic causes oscillations to grow in the LC owing to the negative damping of the characteristic. This then gives a zero DC bias and an exponentially

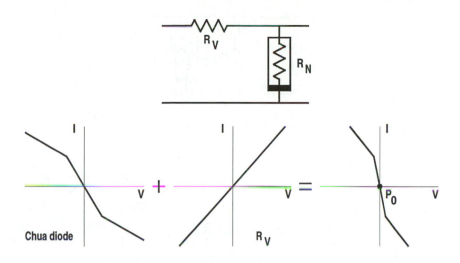

Figure 57.11: *I–V* characteristics for series R_V and R_N.

growing oscillation as shown in the lower trace of Figure 57.8.

Thirdly we consider the voltage across C_1 at point X in the circuit as shown in Figure 57.12. In this case we analyze in terms of the R_V and R_N in parallel and we get the resultant characteristic as shown in Figure 57.12 by adding the currents vertically for each voltage.

This characteristic gives two stable operating points at P_1 and P_2 with an unstable region in between at the origin.

When these three mechanisms are superimposed we see that the voltage across this parallel combination is driven by the amplified output oscillation of the potential divider and that the circuit jumps from one operating point to the other (P_1 to P_2 or P_2 to P_1) as soon as the voltage at X crosses the corners of the characteristic. The energy which has been built up in the oscillations in the LC is transferred to the storage capacitor C_1 and the oscillations in the LC have to start to grow from zero again. This mechanism gives the steps in the waveform across the capacitor C_1 as shown in the upper trace in Figure 57.8 which also has the amplified exponentially growing oscillations superimposed on the steps.

It can also be seen that the value of the capacitance, C_1, is critical. It must be large enough to store the energy built up in the oscillation in the LC circuit but it must not be so large as to smooth totally the output oscillation across C_1 and the potential divider. A value of C_1 such that $9C_1 = C_2$ usually gives satisfactory operation but you should try other values.

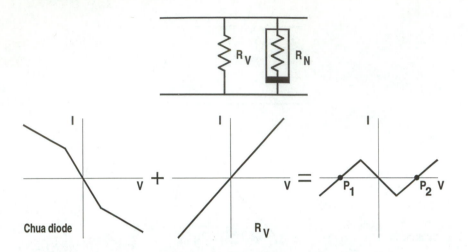

Figure 57.12: *I–V* characteristics for parallel R_V and R_N.

In the discussion of the logistic function we showed that the equation $x_{n+1} = r \times x_n \times (1 - x_n)$ shows a period doubling route to chaos as the parameter r is increased. If the R_V in the Chua circuit is varied slowly and if the circuit is biased towards one of the basins of oscillation by using, say, $5.6\,\text{k}\Omega$ and $3.3\,\text{k}\Omega$ resistors in series with the diodes in Figure 57.4 (a) then a period doubling sequence for the oscillations can be observed and is manifested by a progression from a single closed loop to a double loop to a four fold closed loop when the display is set to XY mode as in Figure 57.9.

In the sinusoidal oscillators which were described in Unit 53, the amplitude was stabilized by the use of a nonlinear feedback component such as a miniature bulb or a thermistor. This gave a stable loop gain of 1 and an oscillator which operated in the linear region. This can be compared with the controlled or limited amplitude oscillations which are obtained with the Chua circuit. In the case of the Chua oscillator, Figures 57.5 and 57.6, the amplitude stabilizes and the value of the voltage at some future time is calculable from $V = V_0 \sin(2\pi ft + \phi)$. In the case of the Chua chaotic oscillator, the loop gain is greater than 1 but the oscillation is prevented from growing indefinitely by the switching action in the circuit which kills a large oscillation and restarts the oscillation. The result is that the voltage at some future time is calculable and deterministic but is not predictable owing to the sensitivity of the system to initial conditions. This is a characteristic feature of chaotic systems.

In naturally occurring systems, the stably, uniformly oscillating system is very rare but systems showing chaotic oscillation similar to the Chua oscillator occur frequently. A leaf fluttering in the wind, a wave at sea, turbulent eddies in the wake of a ship and the beating of a heart are some examples of chaotic systems.

57.1 Problems

57.1 Sketch the I–V characteristic for the circuit in Figure 57.4 (b) if the $300\,\Omega$ resistors are replaced by $22\,\mathrm{k}\Omega$ resistors and the $1.1\,\mathrm{k}\Omega$ resistor is replaced by a $3.3\,\mathrm{k}\Omega$ resistor.

57.2 A pendulum mass is suspended from the top of a rod which is hinged at the bottom at H and is constrained loosely between two limits at the top as shown in Figure 57.13. The pendulum is driven by a pulsed electromagnetic drive which causes the oscillation amplitude to increase with time. Describe the motion of the pendulum bob as a function of time.

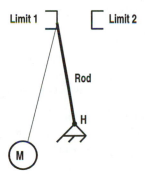

Figure 57.13: Resonantly driven sloppy pendulum.

57.3 Design an electronic circuit which will detect the proximity of the pendulum bob in Problem 57.2 and which will apply a pulse to an electromagnet which is timed so as to increase the amplitude of oscillation of the pendulum.

57.4 A gyrator is an electronic circuit which uses op-amps, resistors and capacitors to simulate an inductor. Figure 57.14 shows the circuit for a Riordan gyrator for which the inductance is given by:

$$Z_{in} = \frac{Z_1 Z_3 Z_5}{Z_2 Z_4}$$

If $Z_1 = Z_3 = Z_4 = Z_5 = 1\,\text{k}\Omega$ and Z_2 is a capacitor of value $C = 0.01\,\mu\text{F}$ having an impedance $Z_2 = \frac{1}{j\omega C}$, calculate the inductance simulated by the gyrator.

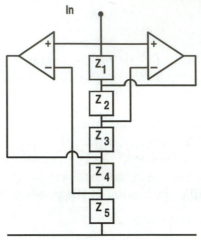

Figure 57.14: Riordan's gyrator simulating an inductance.

57.5 Use the two rules for op-amp operation to derive the relationship:

$$Z_{in} = \frac{Z_1 Z_3 Z_5}{Z_2 Z_4}$$

quoted for the gyrator circuit in Problem 57.4.

57.6 Discuss the characteristics of the gyrator, using the circuit in Figure 57.14, which would result from setting $Z_1 = Z_3 = Z_4 = 1\,\text{k}\Omega$, $Z_2 = 0.01\,\mu\text{F}$ and using a negative resistor circuit for Z_5 of value $Z_5 = -1.5\,\text{k}\Omega$.

57.7 Draw a circuit for a Chua chaotic oscillator which uses a Riordan gyrator in place of the inductor. Calculate suitable values for the components.

Unit 58 Circuit simulation using PSPICE

PSPICE is a simulation program for electronic circuits.
Components are connected between numbered NODES in a circuit.
NODE 0 is taken to be ground.
Resistor labels begin with R.
Capacitor labels begin with C.
Inductor labels begin with L.
Silicon transistor labels begin with Q.
Voltage sources begin with V.
Current sources begin with I.
Subcircuit labels begin with X.
.DC analysis applies a swept DC to the named node.
.AC analysis applies a frequency swept AC to the named node.
.TF calculates the transfer function and Thévenin equivalent.
.TRAN analyzes the transient behaviour over the stated time.
.PROBE prepares an output file suitable for graphing.
.END marks the end of the program.

The original SPICE program was written in Fortran by Laurence Nagle for use on mainframe computers and it enabled electronic circuits to be analyzed prior to construction. Detailed mathematical models for actual circuit components were developed and stored in library files. There are student versions of SPICE available at no charge for use on PCs which have a slightly restricted performance. These PC versions are called PSPICE and can be downloaded from the Internet.

The programs listed in this section were developed and tested using version PSEVAL50 of PSPICE. The programs may require some small modifications if later versions are used. The documentation with the version in use should permit the student to carry out any necessary modifications.

There is only one way to learn to use a computer package and that is to sit at the computer and run examples using the package. These example programs show how some of the electronic examples discussed in the text can be modelled using PSPICE. The effects of substituting different values of the components can be explored. It is also illuminating to investigate how the first order approximations which have been used in this text in the explanations of the circuit operation give descriptions of the circuit operation which are close to the circuit performance calculated with the more complete and detailed library models which can use up to 50 parameters to specify the components.

Since there are many tutorial texts available for PSPICE, the programs will be presented without comment. For convenience, the diagrams are repeated here with the programs. The labelling of the nodes should be readily apparent from the component values and circuit diagrams. These programs should be used as a starting point and the component values and analysis types varied to explore the circuit behaviour.

References:

SPICE for Circuits and Electronics using PSPICE, Muhammad Rashid, Prentice Hall (1994).

SPICE A Guide to Circuit Simulation and Analysis using PSPICE, 3rd edn., Tuinenga P. W., Prentice Hall (1995).

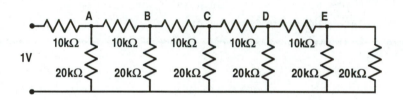

Figure 58.1: Unit 5 Example 5.2.

```
R-2R Ladder
vin 1 0 1
r1a 1 2 10k
r2a 2 0 20k
r1b 2 3 10k
r2b 3 0 20k
r1c 3 4 10k
r2c 4 0 20k
r1d 4 5 10k
r2d 5 0 20k
r1e 5 6 10k
r2e 6 0 10k
.op
.print dc  v(1) v(2) v(3) v(4) v(5) v(6)
.end
```

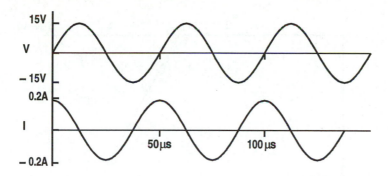

Figure 58.2: Unit 10 Example 10.1 Voltage and current in a capacitor.

```
Unit 10  Example 10.1 Voltage and current in a capacitor
vin 1 0 sin ( 0 15 20kHz )
c1 1 0 0.1uF
.tran 5us 100us
.probe
.end
```

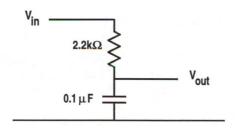

Figure 58.3: Unit 15 Example 15.1 and Unit 16 Low pass filter circuit.

```
Unit 15 Example 15.1 and Unit 16 Low pass filter
vin 1 0 ac 1v
r1 1 2 2200
c1 2 0 0.1uF
.ac  dec 10 10Hz 100kHz
.probe
.end
```

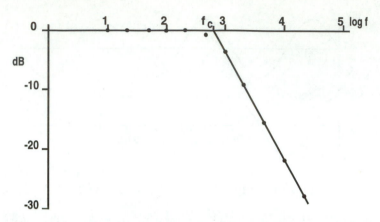

Figure 58.4: Unit 15 Example 15.1 and Unit 16 Low pass filter response.

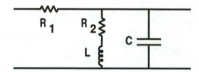

Figure 58.5: Unit 17 Example 17.1 Band pass filter.

```
Unit 17 Example 17.1 Band pass filter
vin 1 0   ac 1V
r1 1 2 1000
r2 2 3 10
L1 3 0 1mH
c1 2 0 0.21uF
.ac dec 20 1kHz 100kHz
.probe  v(2)
.end
```

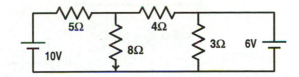

Figure 58.6: Unit 21 Example 21.1 Principle of superposition.

```
Unit 21 Example 21.1 Principle of superposition
va 1 0 10V
vb 3 0 6V
r1 1 2 5
r2 2 0 8
r3 2 3 4
r4 3 0 3
.op
.print dc v(2)  v(3)
.end
```

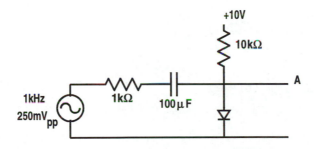

Figure 58.7: Unit 27 Example 27.2 Diode attenuator.

```
Unit  27 Example 21.2 Diode attenuator
vin 1 0 sin ( 0 125mV 1kHz)
r1 1 2 1k
c1 2 3 100uF
d1 3 0 d1N4148
r2 4 3 10k
vdc 4 0 10V
.lib eval.lib
.tran 10us 5ms
.probe v(3)
.end
```

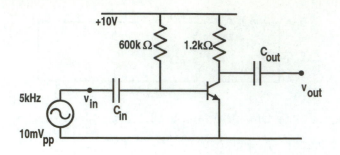

Figure 58.8: Unit 33 Example 33.1 Transistor amplifier.

```
Unit 33 Example 33.1 Transistor amplifier
vcc 1 0 10V
vin 2 0 sin ( 0 5mV 5kHz)
c1 2 3 100uF
r2 1 3 600k
q1 4 3 0 qbc107
r3 1 4 1.2k
c2 4 5 100uF
rload 5 0 1000k
.tran 5us 2ms
.probe
.model qbc107 npn (bf=200)
.end
```

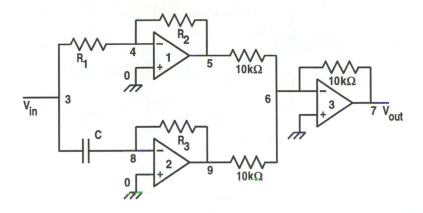

Figure 58.9: Unit 47 Example 47.1 PD controller. PSPICE nodes are numbered.

```
Unit 47 Example 47.1 Proportional plus derivative controller
*Ideal op-amps (Voltage controlled voltage sources) are
*used instead of library ua741 amplifiers.
vin 3 0 pwl ( 0 0 10ms 0  1s 100mv)
r1 3 4 10k
r2 4 5 33k
e1 0 5 0 4 100000
r4 5 6 10k
c1 3 10 1uF
r7 10 8 1000
r3 8 9 2200k
e2 0 9 0 8 100000
r5 9 6 10k
r6 6 7 10k
e3 0 7 0 6 1000000
.tran 5m 1
.probe
.end
```

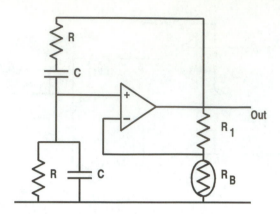

Figure 58.10: Unit 53 Example 53.1 Wein bridge oscillator.

```
Unit 53 Example 53.1 Wein bridge oscillator
vcp 1 0 15
vcn 0 2 15
x1 10 11 1 2 13 ua741
r1 13 4 1000
c1 4 10 1uF
r2 10 0 1000
c2 10 0 1uF
r3 13 11 1000
*Note that Rb is fixed at 400 Ohms
*This gives an exponential growth of the sinusoid waveform
r4 11 0 400
.lib eval.lib
.tran 20us 50ms
.probe
.end
```

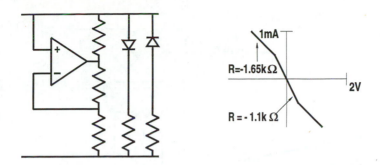

Figure 58.11: Unit 57 The operation of Chua's diode.

```
Chua diode
*negative resistance
vi 4 0  0
r4 4 1 1
r5 1 5 470
r6 5 6 470
r7 6 0 700
v3 10 0 15
v4 0 11 15
x1    1 6 10 11 5 ua741

*nonlinear resistance
r1 1 2 470
d1 2 0 D1N4148
r2 1 3 470
d2 0 3 d1n4148
.dc vi -4 4 .5
.lib eval.lib
.probe
.end
```

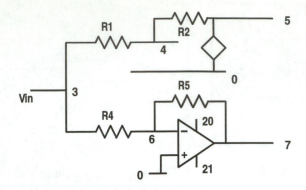

Figure 58.12: Unit 58 Ideal and real op-amps.

```
Test of opamp models
*An Ideal op-amp (voltage controlled voltage source) is
*compared to library ua741 amplifiers
*in inverting amplifier configuration.
vcp 20 0 15
vcn 21 0 -15
vin 3 0 0
r1 3 4 10k
r2 4 5 3300k
e1 0 5 0 4 100000
r4 3 6 10k
r5 6 7 3300k
x1 0 6 20 21 7 ua741
.lib eval.lib
.dc vin -.1 .1 .005
.probe  v(5) v(7)
.end
```

Bibliography

This is a small but representative selection of the many texts which are available and which will provide further reading.

Beards, Peter H. (1991), *Analog and Digital Electronics, 2nd edn,* Prentice Hall.

Fraser, D. A. (1983), *The Physics of Semiconductor Devices,* Oxford University Press.

Hilborn, Robert C. (1994), *Chaos and Nonlinear Dynamics,* Oxford University Press.

Horowitz, P. & Hill, W. (1996), *The Art of Electronics, 2nd edn,* Cambridge University Press.

Jones, M. H. (1985), *A Practical Introduction to Electronic Circuits,* Cambridge University Press.

Kennedy, M. P. (1993), 'Three Steps to Chaos', *IEEE Transactions on Circuits and Systems,* Vol **40** (10), pp 640 and 657, 1993.

Kittel, C. (1961), *Elementary Statistical Physics,* Wiley.

Malvino, A. P. (1993), *Electronic Principles, 5th edn,* Macmillan/McGraw Hill.

Omar, M. Ali (1973), *Elementary Solid State Physics,* Addison Wesley.

Rashid, M. (1995), *SPICE for Circuits and Electronics using PSPICE,* Prentice Hall.

Stauffer, D., Aharony, A. (1992), *Introduction to Percolation Theory 2nd edn,* Taylor & Francis.

Tomlinson, G. H. (1991), *Electrical Networks and Filters,* Prentice Hall.

Tuinenga, P. (1995), *SPICE A Guide to Circuit Simulation and Analysis using PSPICE 3rd edn,* Prentice Hall.

Wilkinson, J. (1994), *An Introduction to Digital Audio,* Focal Press.

Answers

Answers to some (but not all) of the numerical problems are given here as an aid to students. There are many design problems included in the problem sets for which there are no single answers. In these cases a 'correct' answer is one which satisfies the numerical design constraints but which can also be constructed using components with normal tolerances and within the usage limitations specified in the component data sheets.

1.1: 15 mV.

1.2: 12.1 kΩ

1.3: 71.4 kΩ, 64.6 kΩ, 0.31 mA, 0.33 mA, 0.30 mA.

1.4: 9 V, 1.91 mA

1.5: ≤0.96 mA

1.11: 0.25 μm

2.1: 1.88 V, 5.06 V, 7.59 V, 14.53 V

2.2: 3.0 mA, 14.1 V, 14.46 V

2.3: +2.47 V, −3.53 V

3.2: 327 Ω

3.3: 0.89 V, 2.19 mA, 0.16 mA, 0.13 mA

3.4: 7.23 V

3.5: 50 mA, 10 mA

3.6: 0.5 mm^2, 0.64 Ω

4.1: 500 kΩ, 400 kΩ, 50 kΩ, 50 kΩ

4.2: 0.99 V, 0.099 V, 9.9 mV, 0.99 mV

4.4: −9.0 V, −6.54 V, −3.55 V, −0.55 V, +3.73 V, +8.0 V

5.1: 833 Ω

5.2: 800 Ω

5.3: 0.25 V, 62.5 mV, 15.6 mV, 3.9 mV, 0.98 mV

6.1: 2.08 A, 5.76 Ω

6.2: 1.8 Ω, 6.7 A, 33 A

6.3: −900 V, −741 V, −635 V, −529 V, −423 V, −317 V, −211 V, −105 V,

6.4: 11.5 W, 5.7 °C

7.1: −6 dB

7.2: 79 Ω, 24921 Ω

7.3: 22.9 mV, −42.4 dB

7.4: 10^{-6} Wm^{-2}, 1 Wm^{-2}

8.1: 0.269 ms, 0.603 ms, 0.936 ms, 1.269 ms, ..., $T = 0.666$ ms

8.2: 33.9 mV

8.5: 2.84 V, 2.58 V, −5.72 V, 6.69 V

9.1: 2.12 V$_{RMS}$

10.1: 6.9 A$_{RMS}$

10.2: 0.38 A$_{pp}$, 0.565 A$_{pp}$

11.1: 0.28 V

11.2: 1333 As^{-1}

11.3: 5 μH

11.5: 49

12.1: −j1592 Ω, −j19.9 Ω

12.2: +j6.28 Ω, +j3140 Ω

13.2: 2297 Ω, −0.29 rad

324

13.3: $3322\,\Omega$, $+0.116\,\text{rad}$

14.1: $(320 + j1884)\,\Omega$
14.2: $(680 - j48)\,\Omega$
14.3: $(253 - j702)\,\Omega$
14.4: $(11.5 + j2.21)\,\Omega$

15.2: $-1.7\,\text{dB}$, $-0.604\,\text{rad}$
15.3: $-2.5\,\text{dB}$, $+0.724\,\text{rad}$
15.4: $-1.51\,\text{dB}$, $-0.573\,\text{rad}$
15.6: $1856\,\text{Hz}$, $45°$

19.2: $10.93\,\text{V}$, $1277\,\Omega$

20.1: $8.86\,\text{mA}$, $740\,\Omega$

21.1: $19.8\,\text{mA}$
21.3: $8.66\,\text{V}$

22.1: $40\,\text{k}\Omega$, $8.2\,\text{k}\Omega$, $4.0\,\text{k}\Omega$, $559\,\Omega$

23.1: $0.55\,\mu\text{m}$

24.2: $11\,\mu\text{A}$, $16\,\mu\text{A}$, $4.4\,\text{mA}$, $6.6\,\text{mA}$,
$0.5\times10^{-3}\,\text{AV}^{-1}$, $0.22\,\text{AV}^{-1}$

25.1: $0.35\,\text{V}$, $0.44\,\text{V}$
25.4: $0.8\,\text{W}$

26.1: $1.91\,\text{mA}$, $7\,\text{V}$, $0.7\,\text{V}$
26.2: $12\,\text{V}$, $7.48\,\text{V}$, $0.7\,\text{V}$
26.4: $0.977\,\text{A}$, $5\,\text{V}$, $1.09\,\text{V}$

27.2: $4.2\,\text{mV}$
27.3: $11\,\text{mV}_{\text{pp}}$

29.1: $7.4\,\text{V}$, $3.18\,\text{V}$, $0.7\,\text{V}$

29.2: $1.38\,\text{V}$, $1.54\,\text{k}\Omega$

31.1: $0\,\text{V}$, $0.7\,\text{V}$, $6.16\,\text{V}$, $1.16\,\text{mA}$,
$7.75\,\mu\text{A}$, $1.16\,\text{mA}$
31.4: $5.05\,\text{V}$, $5.75\,\text{V}$, $10\,\text{V}$, $1.87\,\text{mA}$,
$6.24\,\mu\text{A}$, $1.87\,\text{mA}$
31.8: $4.2\,\text{V}$, $4.9\,\text{V}$, $9.5\,\text{V}$, $0.75\,\text{mA}$,
$0.75\,\text{mA}/\beta$, $0.75\,\text{mA}$

33.1: -339
33.2: -173
33.3: -71

34.1: $2.82\,\text{V}$

39.3: -17.6
39.7: $12\,\text{k}\Omega$, $216\,\text{k}\Omega$

40.1: $+18.4$

43.1: $-7.05\,\text{V}$, $-0.18\,\text{V}$
43.4: $4.92\,\text{mV}$, $4.1\,\text{V}$

44.1: $0.495\,\text{V}_{\text{amplitude}}$ square wave

45.3: $0.37\,\text{V}_{\text{amplitude}}$ cos waveform

49.1: $27\,\text{dB}$, $\approx20\,\text{kHz}$

50.1: $0.1\,\text{mV}_{\text{RMS}}$

51.1: $0.5\,\text{Hz}$
51.2: $330\,\mu\text{V}_{\text{RMS}}$

55.1: $1.22\,\text{mV}$

Index

UNIVERSITY OF GLAMORGAN · PRIFYSGOL MORGANNWG

Learning Resource Centre

$$V = I \times R$$
$$R_{series} = R_1 + R_2 + R_3 + \cdots$$
$$\frac{1}{R_{parallel}} = \frac{1}{R_1} + \frac{1}{R_2} + \frac{1}{R_3} + \cdots$$
$$C_{parallel} = C_1 + C_2 + C_3 + \cdots$$
$$\text{Power} \quad P = V \times I = I^2 \times R = \frac{V^2}{R} \text{ watts}$$
$$\text{Power ratio} = 10 \log\left(\frac{P_{out}}{P_{in}}\right) = 20 \log\left(\frac{V_{out}}{V_{in}}\right)$$

$$\omega = 2\pi f$$
$$T = \frac{1}{f}$$
$$V = V_0 \sin(2\pi f t + \phi)$$
$$I = I_0 \sin(2\pi f t + \phi)$$
$$V_{Amplitude} = 1.4 \times V_{RMS}$$
$$V_{Peak-to-Peak} = 2 \times V_{Amplitude}$$
$$\text{Air core} \quad L(\text{nH}) = \frac{N^2 d^2}{0.46d + 1.02b}$$
$$\text{Ferrite core} \quad L(\text{nH}) = N^2 A_L$$
$$Z_R = R \quad \text{for resistance}$$
$$Z_C = \frac{1}{j\omega C} \quad \text{for capacitor}$$
$$Z_L = j\omega L \quad \text{for inductor}$$
$$V = ZI$$
$$= |Z| e^{j\phi} I$$
$$\text{or} \quad V_0 e^{j\omega t} = |Z| I_0 e^{j(\omega t + \phi)}$$
$$Z = R + jX$$

where R is the resistance, X is the reactance.
Generalized potential divider:

$$\frac{V_{out}}{V_{in}} = \frac{Z_2}{Z_1 + Z_2} = |A| e^{j\phi}$$

Then $|A| =$ attenuation and $\phi =$ phase shift

$$|c| = |a + jb| = \sqrt{a^2 + b^2} \quad \tan\phi = \frac{b}{a}$$

$$\left|\frac{1}{c}\right| = \left|\frac{1}{a + jb}\right| = \frac{1}{\sqrt{a^2 + b^2}} \quad \tan\phi = \frac{-b}{a}$$

$$f_{corner} = \frac{1}{2\pi CR} \quad \text{or} \quad \frac{R}{2\pi L}$$

Thévenin's theorem $R_{out} = \frac{V_{out\ open\ circuit}}{I_{out\ short\ circuit}}$

Principle of superposition: Replace voltage sources by short circuits and current sources by open circuits.

Semiconductor equation:

$$n \times p = n_i^2 = \text{constant for constant } T$$

Current through a pn diode junction is:

$$I = I_0 \left(\exp\left(\frac{eV}{kT}\right) - 1\right)$$
$$\approx I_0 \exp\left(\frac{V}{25\,\text{mV}}\right)$$
$$\frac{kT}{e} = 25\,\text{mV}$$

Voltage across diode is:

$$V = V_k + I \times R_B$$

where $V_k = 0.7\,\text{V}$ for Si, $0.3\,\text{V}$ for Ge.

$$R_{dyn} = \frac{dV}{dI} = \frac{25\,\text{mV}}{I}$$

Ripple voltage of rectified and smoothed AC is:

$$\frac{I_{out}}{2 \times f \times C}$$

Zener diode conducts in reverse bias when the voltage is greater than the Zener voltage for the diode.

For a bipolar transistor:

$$I_C = \beta \times I_B$$
$$I_C \approx I_E$$
$$V_{BE} \approx 0.7\,\text{V}$$

Basic equation for transistor bias circuits is:

Voltage supply = Sum of individual voltage drops

Common emitter amplifier amplification:

$$A_V = -\frac{I_E}{25\,\text{mV}} \times R_C$$
$$R_{in} = \beta \times \frac{25\,\text{mV}}{I_E}$$

A JFET is specified by:

$$V_{GS(off)} = \text{Gate to source cutoff Voltage}$$
$$I_{DSS} = \text{Drain saturation current}$$
$$g_m = \frac{dI_D}{dV_{GS}} = \text{Mutual conductance}$$
$$I_D = I_{DSS}\left(1 - \frac{V_{GS}}{V_{GS(off)}}\right)^2$$

An enhancement mode MOSFET is specified by:

$$I_D = k(V_{GS} - V_{GS(th)})^2$$

For a JFET common source amplifier select:

- $R_S = \left| \frac{V_{GS(off)}}{I_{DSS}} \right|$

- Then $V_{GS} \approx 0.4 \times V_{GS(off)}$

- and $I_D \approx 0.4 \times I_{DSS}$

- $A_V = -g_m \times R_D$

$$g_m = \frac{dI_D}{dV_{GS}} = -2\frac{I_{DSS}}{V_{GS(off)}} \left(1 - \frac{V_{GS}}{V_{GS(off)}} \right)$$

For op-amps used in linear region:

- **Rule 1.** The voltage difference between the inverting and noninverting inputs is approximately zero.

- **Rule 2.** No current flows into the input terminals of the op-amp.

- The gain of an inverting amplifier is:

$$A_V = -\frac{R_f}{R_{in}}$$

- The gain of a noninverting amplifier is:

$$A_V = 1 + \frac{R_1}{R_2}$$

- $A_V = \frac{1}{\beta} = \frac{1}{\text{Feedback fraction}}$

- The output from an inverting adder is:

$$V_{out} = -R_f \left(\frac{V_1}{R_1} + \frac{V_2}{R_2} + \frac{V_3}{R_3} + \cdots \right)$$

- Output of current to voltage converter is:

$$V_{out} = -I \times R_f$$

- A bridge is in balance when:

$$\frac{R_1}{R_2} = \frac{R_3}{R_4}$$

- For a differential amplifier:

$$V_{out} = \frac{R_2}{R_1} \times (V_2 - V_1)$$

- For a differentiator:

$$V_{out} = -CR_f \frac{dV_{in}}{dt}$$

- For an integrator:

$$V_{out} = -\frac{1}{CR} \int V_{in} dt$$

If a component characteristic is $I = f(V)$ then:

- Putting the component in place of the input resistor of an inverting amplifier gives the forward function:

$$V_{out} = -R \times f(V_{in})$$

- Putting the component in place of the feedback resistor of an inverting amplifier gives the inverse function:

$$V_{out} = -f^{-1} \left(\frac{V_{in}}{R} \right)$$

Frequency response of a 741 op-amp:

- Has a corner at 10 Hz and 100 dB.

- Open loop gain decreases by 20 dB per decade above 10 Hz

Noise in bandwidth B has units of:

Volts per $\sqrt{\text{Hz}}$ or Amps per $\sqrt{\text{Hz}}$

- Thermal noise from a resistor, R, at temperature T within a bandwidth B is:

$$V_{noise} = \sqrt{4kTRB}$$

- The shot noise is:

$$I_{noise} = \sqrt{2eIB}$$

- Flicker noise spectrum varies as

$$\frac{1}{f} = \frac{1}{\text{Frequency}}$$

For a 555 Timer IC:

$$T_1 = 0.7(R_A + R_B)C \quad \text{and} \quad T_2 = 0.7R_BC$$

Sinusoidal voltage waveforms are obtained by using an amplifier with:

- Positive feedback,

- Loop gain of 1

- Frequency selective feedback network.

An R–$2R$ ladder gives an output:

$$V_{out} = -R_f \frac{V_{ref}}{4R} \left(S_0 + \frac{S_1}{2} + \frac{S_2}{4} + \frac{S_3}{8} + \cdots \right)$$